The Ecology of Sex

Paul J. Greenwood,
Department of Adult and Continuing Education,
University of Durham.

and

Jonathan Adams,
Department of Adult and Continuing Education,
University of Leeds.

Edward Arnold

First published in Great Britain 1987 by
Edward Arnold (Publishers) Ltd, 41 Bedford Square, London WC1B 3DQ

Edward Arnold (Australia) Pty Ltd, 80 Waverley Road, Caulfield East,
Victoria 3145, Australia

Edward Arnold, 3 East Read Street, Baltimore, Maryland 21202, U.S.A.

British Library Cataloguing in Publication Data

Greenwood, Paul J.
The ecology of sex.
1. Sex (Biology) 2. Evolution
I. Title II. Adams, Jonathan
574.1′66 QH481

ISBN 0-7131-2934-4

Text set in 10/11 pt Times Compugraphic
by Colset Private Limited, Singapore
Made and printed in Great Britain
by Richard Clay Ltd., Bungay, Suffolk

Preface

This is a book about sex; it is not a book about reproduction. To ensure their genetic survival all organisms strive to reproduce, to form new individuals as effectively as possible. But they could do this by asexual or sexual means. Sex and reproduction are not synonymous, although the distinction is often confused.

An example may help to clarify our intentions. Consider that ubiquitous woodland bird and ecological paradigm, the great tit *Parus major*. Detailed studies have been made of this bird's reproductive strategy. Ecological factors lead to adaptive variations in the size of clutch a pair produces; in their investment in hatchlings; and in their fledglings' success. This book is not about that strategy: it is about all the problems that precede egg-laying. Why do great tits reproduce sexually and why are there two separate sexes? Why should sex be determined by the fusion of particular gametes? Why do males sing and defend a territory? And why are male great tits monogamous?

There are many excellent books about behaviour associated with sex, particularly courtship in birds and mammals. Consequently we only touch on this area of behavioural ecology and look instead towards evolutionary ecology, which has received less attention. We have necessarily been selective in choosing examples to illustrate concepts, but we think that these studies are some of the best of a much broader spectrum. The reader should remember, however, that all the theoretical rules we discuss apply equally to plants, to invertebrate animals and to vertebrates. Taxonomic biases exist in our examples simply because some theories have only been tested with some groups.

The line drawings of animals and plants which complement the text are the work of Diane Breeze. We are grateful to Paul Harvey for his detailed comments; to Tony Holley, Caroline Naylor, Ian Swingland and Peter Wheeler for discussion about specific sections; and to those students on our adult education courses who have encouraged us to write this book.

Durham — P.J.G.
Leeds — J.A.
1987

Contents

1 Sex and Sexes

Natural selection shapes the course of evolution. It acts on the heritable differences that occur between individuals. Those that are better suited to the rigours of life will survive and reproduce. Those that are less well adapted will die out. Ultimately, new species will evolve and others will become extinct.

Sexual reproduction involves the fusion of gametes and the recombination of genetic material. The diversity and complexity of life owes much to the evolution of sex. But reproduction does not depend on sex and the two should not be confused. Many plants, such as strawberries, and animals, such as corals, can propagate themselves without any form of sex. It is no coincidence, however, that groups such as birds and mammals rely exclusively on sexual reproduction. Without sex such complex organisms would never have evolved through natural selection. Yet the primitive mode of reproduction must have been asexual: no sex, no sexes. Why did sex evolve?

1.1 The evolution of sex

Perhaps the easiest way of approaching the problem of the origin of sex is to look briefly at the two main types of cell division: mitosis and meiosis. Imagine a unicellular organism, such as *Amoeba*, with a diploid (2n) number of chromosomes. Through the process of duplication and division that occurs during ***mitosis*** the full complement of chromosomes is passed to each daughter cell. If there are a number of variations that arise in the genetic make-up of the original cells then, in the absence of further mutation, those variations will be faithfully, transmitted to the next generation. The production of two identical individuals from one parental cell is a form of asexual reproduction.

During ***meiotic*** division a diploid cell undergoes a reduction to produce ***gametes***, each with a haploid (n) number of chromosomes. But recombination also occurs during meiosis, when genetic material is exchanged between homologous pairs of chromosomes. This has the potential to create an enormous number of varieties from one generation to another. A new individual is subsequently formed by the fusion of two gametes from two parental individuals. This process of meiotic division, recombination and gametic fusion is ***sexual reproduction***.

The earliest forms of cellular life must have been haploid and reproduced, asexually, by growth, replication and division. Nowadays, most species of plants and animals reproduce sexually. It is unclear when sexual reproduction first evolved; it could have been anytime between three and one billion years ago. One possible reason for its evolution is that there might have been a selective advantage for primitive haploid cells to fuse together. Damage that occurred to the genetic material (DNA) of one cell could be masked by the equivalent but undamaged piece from another cell. But there is a further problem. Even when sex evolved it is not clear how it was maintained against re-invasion by its precursor: asexual reproduction. This is an important evolutionary question. At first sight there appear to be considerable individual advantages to asexual reproduction and considerable costs to sex.

1.2 The costs of sex

Asexual reproduction of single-celled species, such as protists and blue-green algae, involves duplication of chromosomes followed by a division. Sexual reproduction, where gametes of equal size are produced (***isogamy***, see Fig. 1.2), requires the several processes of meiosis, duplication, recombination, two reduction divisions and, finally, fusion. Thus meiosis is likely to be energetically more expensive and more time consuming than mitosis. In addition recombination has the potential to break up those genetic arrangements that are well adapted to the environment.

Further costs arise in those species which have sexes and two sorts of gametes. This condition, known as ***anisogamy*** (see Section 1.5) refers to those species where some individuals (males) produce small gametes, usually called sperm or pollen, and other individuals (females) produce large gametes such as eggs or ovules (Fig. 1.1). Females produce, on average, equal numbers of both sexes (Chapter 4). But, if a male makes no other contribution to the rearing of offspring than sperm, then a female that produced eggs which developed directly into daughters without the need for fertilisation (a condition called ***parthenogenesis***) would have a numerical advantage. For example, females rearing two productive daughters each generation by parthenogenesis would eventually swamp those that produced, typically, one daughter and an unproductive son.

There could be an advantage to parthenogenesis even in those species where males do make some parental investment and, as a result, a female raises twice as many young as she could rear alone. Imagine a mutant female which produced diploid eggs that developed into offspring without fertilisation. She could mate with a male and accept his assistance, yet the offspring would receive all her genetic material instead of half of hers and

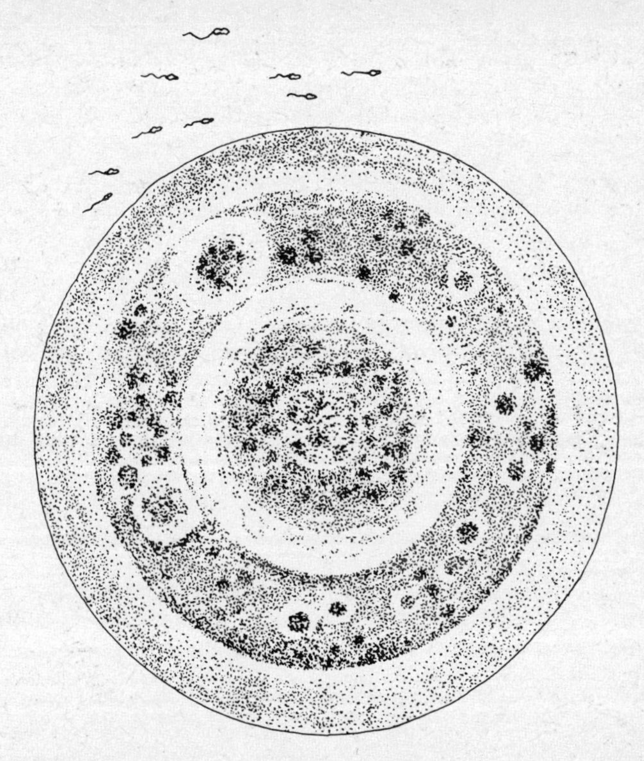

Fig. 1.1 Human egg and sperm. The size difference between the male and female gametes among humans is typical of ***anisogamous*** species. Sperm are much smaller, and cost much less to produce. They can be produced in large numbers whereas eggs are a relatively scarce resource.

half her mate's. Thus, in the short term, she would gain an advantage over normal sexual females.

A final cost to sex concerns the elaborate paraphernalia that accompanies its execution. Behaviours geared exclusively to obtaining a mate (Chapter 5), such as fighting and courtship, and the production of elaborate weapons and gaudy plumage are costs which asexual species do not have to pay.

1.3 The advantages of sex

Asexual reproduction appears to be both relatively more efficient and more productive than sexual reproduction. What, then, are the factors that might account for the success of sex in the evolutionary race? A number of explanations have been put forward, none of which has received universal approval. The main contenders and their problems will be briefly described here.

When mutations occur they are, on average, more likely to be harmful than beneficial. The daughter cells of an asexual line will inevitably inherit all such mutations. Over succeeding generations there will be an accumulation of more and more slightly deleterious characters. Species with sexual reproduction, on the other hand, have the potential to reduce their genetic liability through recombination. Mutations could, by chance, be off-loaded during recombination onto one set of chromosomes rather than the other. Among the variety of new combinations that are produced there will be some individuals which will then inherit a surfeit of genetic defects. But there will also be some individuals, produced from gametes carrying fewer mutations, which will have a smaller genetic load than their parents. These offspring will be better adapted to cope with their environment than any asexual progeny.

There are two other consequences of sexual reproduction. First, the fusion of gametes from different parental stocks could bring together favourable combinations of genetic material. It must be borne in mind, however, that sex can also disrupt such combinations. Second, sex generates the variation for natural selection to work on. As a result sexual species should evolve at a much faster rate than asexual species in response to pressures such as competition and predation. Asexual species should be more likely to go extinct because of their lack of genetic flexibility.

The benefits of sexual reproduction presented thus far are frequently interpreted as long term ones which accrue to the population or species. Biologists are often reluctant to accept explanations for evolutionary events which rely on group, rather than individual, selection. Alternative hypotheses for the maintenance of sex have been put forward which stress the short term advantages of sex to the individual.

Assume that the environment is patchy or unpredictable and that it changes from one generation to the next. The variable offspring produced by sex may, by chance, include individuals with genotypes suited to the new patches. As we have seen, asexual reproduction cannot generate that degree of variation. The analogy first drawn by G. C. Williams and repeated many times since is to that of the raffle. Asexual offspring entering a new patch will all have the same ticket number. Sexually produced offspring will each have a different number. The probability that an asexual individual wins is much smaller than that for a sexual type. When they do win they can dominate the patch but with identical numbers these winners must fight over the same prize. With their range of numbers the sexual types are more likely to be represented by an outright winner and several runners-up. But competition is less intense because the prize pool is larger and can be shared.

An organism may experience an unpredictable patch in many different ways. It could be a change in abiotic factors such as climate. Perhaps more important in the short term are biological enemies such as harmful bacteria and viruses, other parasites and predators. A novel virulent strain that successfully infects one asexual individual would inevitably be able to infect others of the same clone. Sexual reproduction results in a range of genotypes which may, by chance, include some which are resistant to infection.

1.4 Ecological correlates of sex

The origins and maintenance of sex are still a matter of debate. Some insights may be gained by comparing the patterns of sexuality and asexuality in the Plant and Animal Kingdoms. Asexuality is probably commoner among plants than animals. Most parthenogenetic strains are almost certainly derived from sexual forebears. They occur relatively infrequently and tend not to replace their sexual ancestors throughout their range. This suggests, firstly, that such species are subject to a high rate of extinction and, secondly, that when they do arise they are comparatively unsuccessful in competition with sexual forms.

Unpredictable and complex environments should favour sex while simple communities with greater predictability and fewer predators or pathogens should have a higher frequency of asexual species. This appears to be the case. Biologically complex habitats such as tropical rain forests with a myriad of interspecific interactions have relatively few asexual species. Simpler or impoverished habitats, such as those at high altitudes and high latitudes, tend to have relatively more.

Asexual offspring often develop quickly and in association with their parent in the same environment. One reason to reproduce sexually is that the offspring may undergo a phase of dispersal into a patch which is novel and unpredictable. To illustrate this pattern further it is possible to compare the timing of reproduction in those species which have both sexual and asexual phases in their life cycle. In other words, the sexual alternation is evolutionarily stable and not transitory. It is therefore presumed that both phases when they occur must confer some advantage.

During the summer, female greenfly (Aphidoidea) which are common garden pests reproduce parthenogenetically at high rates. At this time males are absent or rare in the population because the female aphid is able to control her reproductive cycle and continuously produce diploid eggs hatching directly into daughters. Later in the season the females produce sons.

The cloning process allows a genetically fit female to saturate a habitat with her offspring in a short time under the favourable conditions of summer. When conditions deteriorate in the autumn, however, the advantages of rapid proliferation and genotypic conservation become outweighed. Males can recombine their mother's genes with those of other females and produce a variety of phenotypes. It is likely that one of this variety will be more fit to survive the winter than would the original female.

The small water-flea *Daphnia* has a similar change from asexual cloning to sexual reproduction. This is triggered not only by seasonal change but also by local environmental conditions experienced by the mother. After the production of males the mothers produce special ephippial eggs (enclosed within a protective membrane system called the ephippium). These eggs develop only if fertilised by a male. They are resistant to desiccation and freezing and so are ideal for enduring winter or drought conditions, although they can develop immediately.

1.5 The evolution of sexes

When sex first evolved in eukaryotic, unicellular species it is almost certain that the gametes that were produced were similar in size and probably small. A familiar example of a species with isogamy is the filamentous green alga *Spirogyra* (Fig. 1.2). But, as noted in Section 1.2, the vast majority of plants and animals produce two sorts of gametes: a small motile gamete which fuses with a large gamete with food reserves. Both isogamy and heterogamy can be found in different species of *Chlamydomonas*, the unicellular green alga. Why did male and female gametes evolve?

As animals and plants increased in complexity there may have been selection for an increase in gamete size. A zygote formed from the fusion of two large, well nourished gametes may perhaps have had a better chance of survival and growth into a reproductive organism. But a large gamete may also have had some drawbacks.

Sexual reproduction will have evolved initially amongst species living in an aquatic medium. Large gametes are likely to be less manoeuvrable and therefore less likely to encounter another gamete. Under these circumstances the system would have been open to invasion by an alternative type: one that was small and motile and could selectively seek out (parasitise) the largest available gametes. The outcome would be disruptive

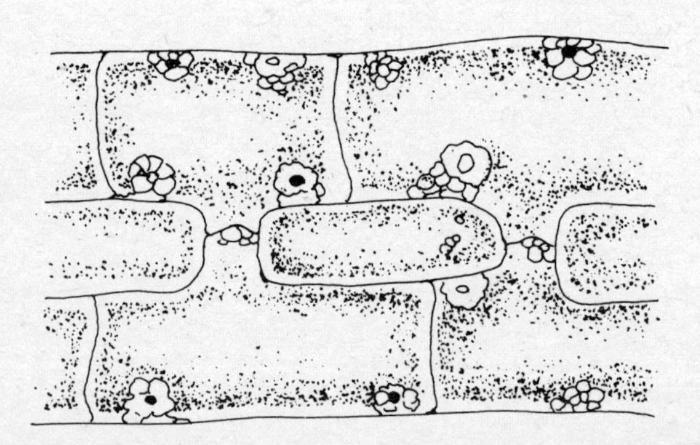

Fig. 1.2 Reproduction in *Spirogyra*. This abundant freshwater alga is typical of species that reproduce sexually but where there is no distinct difference between the conjugating partners. This is ***isogamous*** reproduction.

selection. Medium sized gametes would be neither sufficiently mobile to out-compete smaller ones nor have enough food reserves to produce viable zygotes when fertilised by them. Thus the stage was set for the appearance of two distinct types within a population: male and female sexes.

2 Sex Determination

2.1 Sex determination and sex chromosomes

In humans it is easy to recognise a distinct pair of chromosomes which are the bodies responsible for determining the sex of their carrier. All the other chromosomes, or autosomes, can be matched into morphologically similar pairs in both sexes. In one sex, the human male, the two sex chromosomes differ in appearance. They are heteromorphic (XY) and males are called the heterogametic sex. Females have a homomorphic (XX) pair of chromosomes and are the homogametic sex. The male's XY pair segregate at meiosis and sperm carry either an X or Y chromosome. Although the sex chromosomes do actually resemble these letters, the fact that they are so named is coincidental. In 1891 the X chromosome was first seen under a microscope as a densely staining round body, and named by Henking as 'X', the unknown factor. In eggs fertilised by Y sperm the gonads subsequently develop as testes, but they will otherwise develop as ovaries. It thus appears that the mammalian Y chromosomes is male determining, whereas the number of X chromosomes has no major effect on sex determination. XO mice are female while XXY mice are male.

Among birds, females are the heterogametic sex and the X/Y convention is altered to Z/W. Female birds are genetically ZW and males are ZZ. Among reptiles there is a wide range of conditions and heteromorphic sex chromosomes occur occasionally in turtles, tortoises and some lizards. Those snakes which have heteromorphic sex chromosomes may be the more recently evolved families such as the Viperidae: the heterogametic sex is usually female. Among fish, heteromorphic sex chromosomes are usually absent but the presence of sex determining chromosomes has nonetheless been inferred from the presence of sex-linked genes which appear to act as markers. It should be pointed out, however, that characters expressed in just one sex are sex-linked only in that sense. They can be switched on by the development of one type of gonad rather than the other but the genes themselves may be present on autosomes in both sexes.

2.2 Haplodiploidy

In some species sex is not determined by single chromosomal differences. The production of male greenfly can be controlled by the mother during chromosome segregation at egg production. In other species, including the Hymenoptera (wasps, bees and ants), males develop from unfertilised eggs which carry only half the female chromosomal number. This mode of sex determination is referred to as ***arrhenotoky*** or sometimes as haplodiploidy because of the contrast in the number of chromosomes. Where females are the haploid sex the phenomenon is called ***thelytoky***. Haplodiploidy permits females to control the sex and sex ratio of their offspring. This may be adaptively significant.

When species of parasitoid wasps are presented with a range of host sizes, females lay diploid eggs in larger hosts which turn into females, and haploid eggs which develop as males in the smaller hosts. An example is the ichneumonid wasp *Pimpla* which attacks the pupae of Lepidoptera. When offered a mixture of large (hawkmoth, *Sphinx*) and small (cabbage white, *Pieris*) pupae, *Pimpla* choose to lay female eggs in *Sphinx* and male eggs in *Pieris*. Adult female parasitoid wasps are usually much larger than males of the same species and this pattern of egg-laying behaviour probably matches the nutritional requirements of the larvae to the available hosts.

What happens when a wasp has only one size class of hosts from which to choose? *Coccygonimus* females initially base decisions on former experience, continuing to lay female eggs if hosts are relatively large. As time goes on, however, the sex bias shifts towards males even where host size is unchanged. Such modification is only feasible under parental control in this type of sex-determining system. In these examples the sex of the offspring has been matched at conception to the nutritional state of its environment by the mother. In other cases it may be more appropriate for sex to be determined at a later stage in development.

Haplodiploidy could have arisen first among parasitic Hymenoptera but has probably been a subsequent major factor in the evolution of the social insects. Sons and daughters have different degrees of relatedness to one another and to their mother. Founding queens in a colony mate during a nuptial flight with, usually, one male. Sperm are stored in a spermatheca for subsequent use and the queen regulates the production of fertilised eggs (which develop as workers or new queens) or unfertilised eggs (males). Daughters inherit half their genes from the queen but sons inherit all their genes from their mother. Daughters also inherit all their father's genes. So, on average, female offspring are more closely related to their sisters than they are to any other group (Fig. 2.1). These females might then increase their genetic representation in future colonies by helping to raise more sisters than by becoming mothers themselves. This is an example of a process known as kin selection.

One conflict in this system is that females should also try to raise their own sons. The formation of castes can control this possibility and also

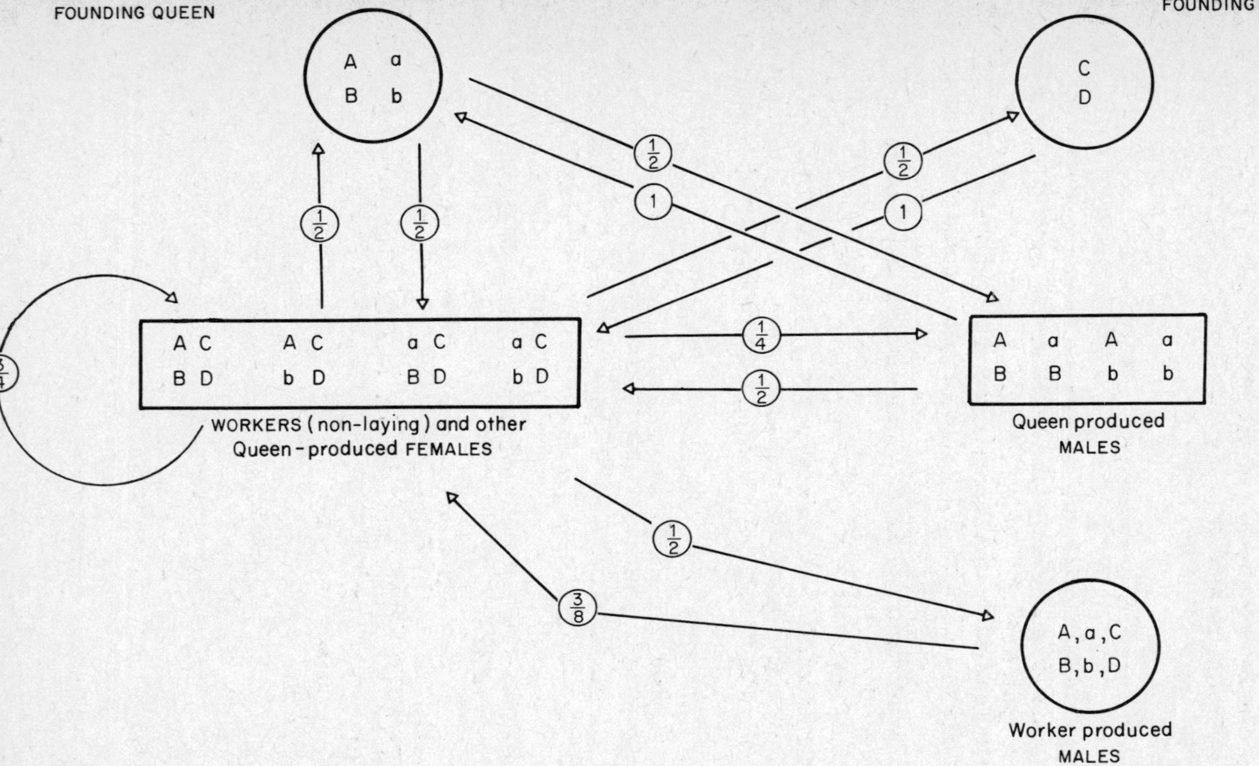

Fig. 2.1 Hymenoptera are haplodiploid insects. Males develop from unfertilised eggs and, in social species, the founding queen can choose which eggs will be fertilised. The figure illustrates the transmission of three alleles at each of two loci (A, a, C and B, b, D) to show how this system of sex determination affects the average degree of relatedness within a colony. Relatedness is indicated by the small figures within circles; female workers are more closely related to one another than to any males produced in the colony. Adapted from Figs. 3.1 and 3.2 in Oster, G.F. and Wilson, E.O. (1978). *Caste and Ecology in the Social Insects*. Princeton

improves the overall efficiency of the colony. Daughter-workers may be nutritionally inhibited from becoming reproductive. This occurs in many ant species, where workers lack ovaries. In other social species males may be produced by both queen and workers, especially when the queen becomes older and less dominant.

2.3 Environmental sex determination

In some organisms the sex of the zygote is not determined at conception. The sex of the individual is instead determined at some point later in its life cycle in response to an environmental cue such as light, temperature or local feeding conditions. Some well documented cases of environmental sex determination (ESD) are listed in Table 2.1. It should be noted that while ESD is often compared with genetic sex determination, inferring heterogamety or haplodiploidy, it can still be an adaptive mechanism under genetic control and subject to evolutionary modification.

When sex is determined at conception, sexual differentiation can start early in development. With ESD, differentiation is delayed and occurs later in the life cycle. What are the conditions which are likely to favour ESD and this delay? Consider, for example, a species in which fitness is

Table 2.1 Examples of environmental sex determination.

Species	*Environmental cue*	*Effect*
Desert shrubs	Soil moisture	Dry conditions increase % males
Potato root eelworm *Heterodera rostochiensis*	Nutrient availability	Poor conditions increase % males
Echiura *Bonellia viridis*	Presence of female in substrate	Future larvae usually male
Crustacea, Isopoda *Ione thoracica*	Presence of female on host	Further parasites usually male
Crustacea, Amphipoda *Gammarus duebeni*	Day length	Males develop early in season
Pisces *Menidia menidia*	Sea temperature	Females develop early in season
Reptilia *Alligator alligator*	Temperature	High temperatures increase % males
Testudo graeca	Temperature	High temperatures decrease % males
Chelydra serpentina	Temperature	Males are produced between 22–26°C, females are produced above and below these limits

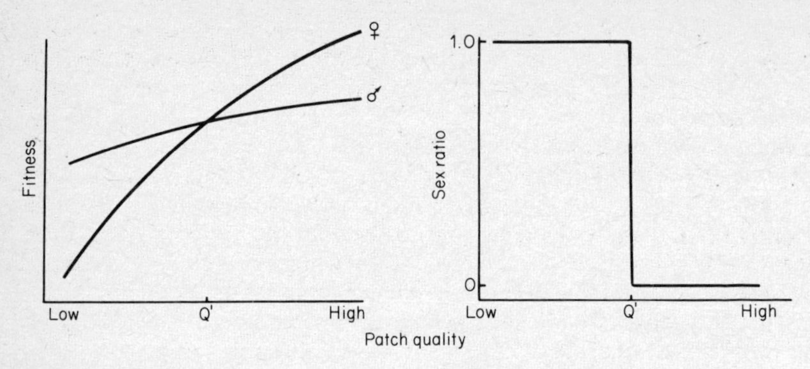

Fig. 2.2 In patchy habitats sex determination may be cued by an environmental factor. Fitness of both males and females is affected by the quality of the patch type into which they are born. In this example, female fitness is more strongly related to patch quality than male fitness; conversely, poor females suffer more compared to other females. If patch quality is predictably related to an environmental feature which the organism can monitor then sex ratio (proportion males) should change from all male below some critical level (Q′) to all female above. Adapted from Fig. 10.B in Bull, J.J. (1983). *Evolution of Sex Determining Mechanisms*. Benjamin/Cummings, Menlo Park, California.

size-related and where environmental patchiness affects the growth potential of an immature animal. In Fig. 2.2 a general model is illustrated where both male and female fitness increase with size. Here, small males are only slightly less fit than large males, but small females have drastically reduced fitness compared to large females. Large size is more advantageous and small size more deleterious to females compared to other individuals of the same sex. In these circumstances an immature animal which entered a patch which is likely to permit only restricted growth should become a female: the sex which suffers least under these conditions. Generally ESD will be favoured for three reasons. First, when the offspring find themselves in an environment which contains a variety of patches with qualities that affect males and females in different ways; secondly, where the quality of the patch is a good indicator of future reproductive expectations for an adult of either sex; and thirdly, where the nature of a particular patch cannot be predicted by the mother.

The nature of environmental patchiness may be of several different kinds. Growth potential can be affected by temperature, food availability, or simply available time. Many species of parasitic Crustacea, for instance, have environmental sex determination. The first individual to arrive on a host develops as a female and grows to a large size. Subsequent arrivals develop as males and remain small. Here, each larva cannot predict before settling whether its host will already be occupied. There is potential for only one large female on a host.

Stegophryxus is an isopod which parasitises the hermit crab, *Pagurus*.

Late larval stage individuals attach to the host's abdomen and develop into females. Larvae settling on females already *in situ* on the host develop into males. The male is tiny compared to his mate and spends the rest of his life inside her brood pouch. A removal experiment tests whether sex is genetically or environmentally determined. A presumptive female *Stegophryxus* which lands directly on the host can be removed and placed on another female. They then differentiate as males. Presumably a factor determining maleness is taken in with food from the female. *Bonellia viridis* is a free-living echiuran marine worm which has a similar sex determining system. Females live in burrows and can be up to one metre in length. They consist of a trunk and large proboscis. Larvae which land on this proboscis develop as males and grow to only a few millimetres. This extraordinary contrast is probably the most extreme case of sexual size dimorphism in the Animal Kingdom (see Fig. 5.1).

Many nematodes invade plant and animal hosts during their larval stages. At low infection rates the adults recovered from a host are predominantly female. At high infection rates the adults are usually male (Table 2.2). It has been suggested that there is a change in sex determination due to crowding at high densities. This could be mediated by an hormonal effect but it is likely that food availability and growth potential also play some role.

Table 2.2 Sex determination in Nematoda. Mermithid nematodes parasitise a variety of plant and animal hosts. The sex ratio of emerging larvae depends on crowding in the host (mosquito larvae). Sex ratios are male biased in smaller hosts and at high densities. (From Bull, J.J. (1983). *Evolution of Sex Determining Mechanisms*. Benjamin/Cummings Publishing Co.)

	Sex Ratio	
Nematodes per host	Large host *Psorophora confinnis*	Small host *Uranotaenia sapphirina*
1	0.0	0.4
2	0.2	0.75
3	0.3	0.9
4	0.45	0.8
5	0.5	0.95

Among plants female fitness in relation to seed production may increase more rapidly with size than does male fitness in terms of pollen. If parent plants release huge numbers of seeds that are scattered in unpredictable patches nearby then those seedlings which germinate in poor soils have only a limited growth potential. In species with separate sexes these should develop as males. Seedlings in good soil should develop as females. Such patchiness may account for the distribution of silver maple, *Acer saccharinum*, in Michigan where trees are clumped on the basis of size and sex. This pattern may be the product of ESD, where poor nutritional conditions favour males and good conditions favour females. However, this pattern could also be caused by sex-related differential mortality if, for instance, female seeds are present but fail to germinate under poor conditions. Since a large proportion of seeds often fail to germinate further work is necessary to confirm the presence or absence of ESD in such species.

A contrasting example is offered by the brackish water shrimp *Gammarus duebeni*. During the breeding season *Gammarus* form mate-guarding pairs in which a large male holds a smaller female until she is available for copulation. A female's reproductive fitness is a function of the number of eggs she can produce, which increases with size. A male's fitness is a function of the range of females he is able to carry. Whereas large males are able to carry the largest and most fecund females, small males are unable to carry a mate at all. The breeding season is restricted in

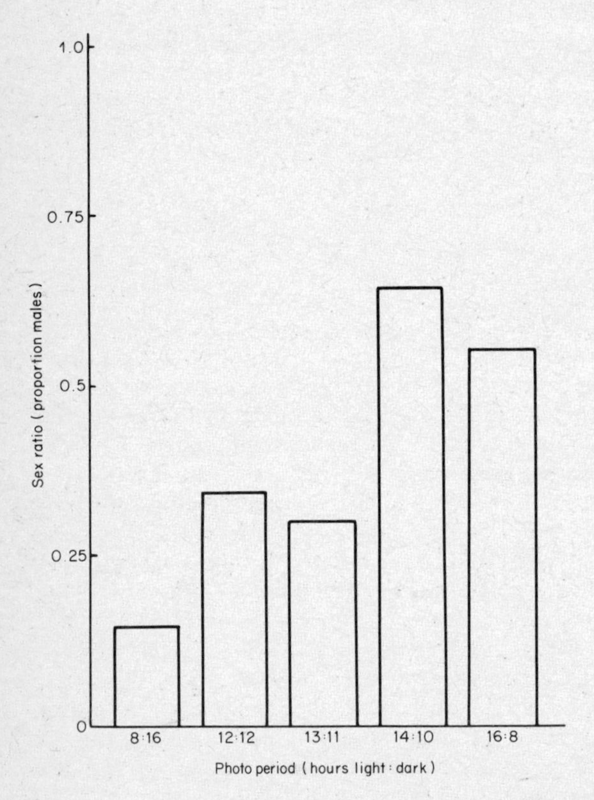

Fig. 2.3 *Gammarus duebeni* is a brackish water shrimp which has a sex determining system cued by daylength. Groups of shrimps from the same brood are incubated at a range of daylengths and the proportions of males and females are recorded. More males develop under longer daylength regimes. (Redrawn from Bulnheim, H.P. (1978). *Helgolander wissenschaftliches Meeresunters*, **31**, 1–33.)

G. duebeni and generations do not overlap in northern Europe. Males which are small during the breeding season have very low fitness. The size of adults is related to the time available for growth in the previous year and the daylength experienced by the juveniles is a good index of their growth potential. Thus offspring which are born in spring will have the opportunity to grow throughout the summer. They have the potential to be large individuals in the following year's breeding season and should differentiate as males. Those born later in the year, which are still small during the short day periods in the autumn, have only a limited growth potential and should develop as females.

Experiments in cabinets with constant temperature and controlled daylength have shown that photoperiod is indeed a cue for sex determination in *G. duebeni*. In short daylength conditions (photoperiod 8 hours light: 16 hours dark) animals from broods which have been split at hatching are more likely to be female. Their siblings which have been kept in a long photoperiod develop as males (Fig. 2.3). In this example the environmental patches are time periods. The individual cannot predict before hatching the amount of time that will be available for growth. Hence it is only at a later stage that it becomes clear what its future reproductive fitness might be. In some other species of *Gammarus*, where generations overlap, sex does not appear to be environmentally determined.

2.4 Sex determination in reptiles

While many snakes and some other reptiles do have an heterogametic Z/W sex determining system others, such as turtles, tortoises and some lizards, have environmental sex determination. One example is *Alligator mississippiensis*. The sex of the alligator is determined by the temperature at which the egg has been incubated. At higher temperatures males are produced and at lower temperatures all eggs hatch as females. The 'unknown' patchiness in the environment is not temperature, however, since this is readily apparent to the parents. Indeed, sites with specific temperature regimes could be selected by the mother. By choosing to lay eggs on levees, which are hot, she can select a male biased sex ratio whilst wet marshes are consistently more likely to produce a female biased sex ratio (Fig. 2.4).

The adaptiveness of this system is much less clear than for other groups with ESD. It may lie in the ability of the parents to respond to fluctuations or perturbations in the sex ratio of the adult population (Section 4.3). If the population were male biased then the fitness of hatchlings would be enhanced if they were female, the rarer sex. Mothers able to monitor the population sex ratio could then select appropriate nest sites. But there is a problem with this argument. Alligators, and other long-lived reptiles such as turtles, have a prolonged maturation period. This means that the sex ratio of the population at the time of egg laying is of little predictive value in relation to the sex ratio when the hatchlings have grown up. An alterna-

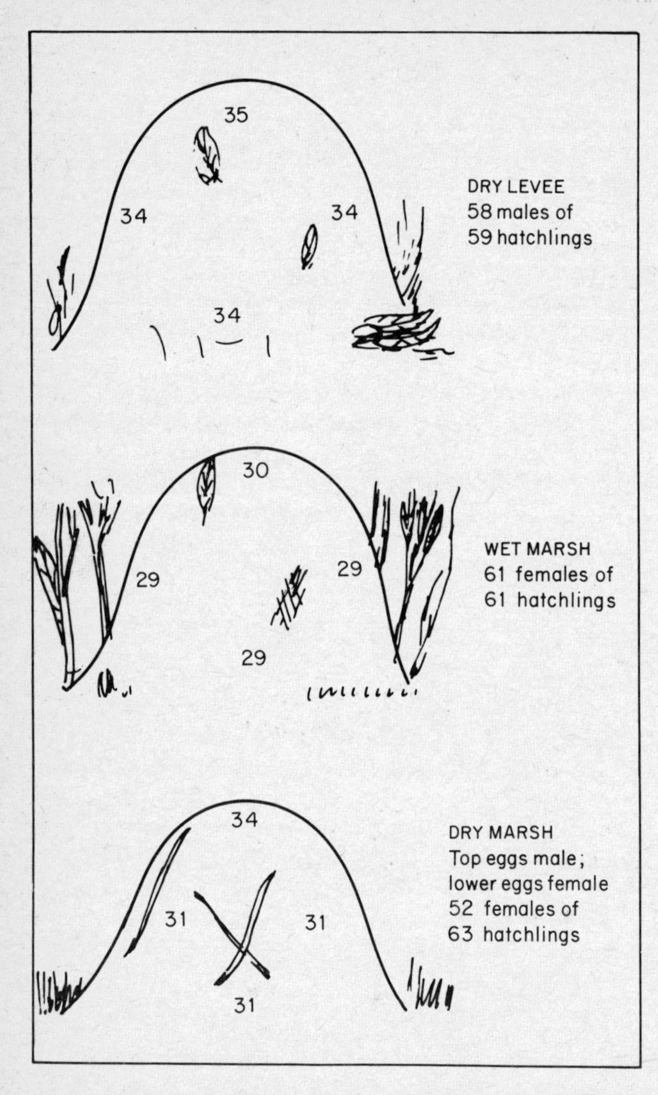

Fig. 2.4 North American alligators have a sex determining system cued by temperature. Alligators lay their eggs in nests, which are mounds of vegetation about 1.5 m diameter and 1 m high. Nest temperature (shown here in °C) varies according to the habitat and is usually hotter higher in the mound. In Louisiana, marsh nests are more common than levee nests and the overall sex ratio is about 5 females to every male. (From Ferguson, M.W.J. and Joanen, T. (1982). *Nature*, **296**, 850–3.)

tive possibility is that opportunities for other modes of sex determination have never arisen and that ESD is the evolutionary consequence of this constraint.

The existence of temperature-dependent sex determination among the turtles and tortoises has major implications for those people concerned with their conservation. One way of safeguarding endangered turtles is to ensure that their eggs have the best chance of hatching and that young turtles safely reach the open sea. An obvious way of doing this is to monitor egg laying of adults and retrieve their clutches before they are eaten or stolen. Eggs can then be safely hatched in battery incubators. It will be obvious, however, that this type of incubator is designed to maintain a constant environment. Where the species has temperature-dependent sex determination the eggs will, of course, all develop as one sex. This problem will not be apparent immediately as the immature turtles are not easily sexed. However, the release of a single sex batch of hatchlings will be of little benefit to the population, particularly if they are all male, and the work put into the conservation programme is thus wasted.

3 Sex Allocation

3.1 Sex allocation

Among a very large number of species of plants and animals individuals are either male or female. These are referred to as dioecious or ***gonochoristic*** species. But there are also species in which individuals function as both male and female during some part of their life. Sometimes they function as both sexes at more or less the same time, in which case they are called ***simultaneous hermaphrodites***. In other cases they function as one sex for part of their life changing to the other sex later, or they may switch back and forth several times in successive reproductive periods. These are sequential hermaphrodites or ***sex-changers***.

In this chapter the general conditions which favour hermaphroditism over dioecy will be discussed, and both costs and benefits will be illustrated with examples among plants and animals. One consequence for individuals which are simultaneously male and female is that they have the potential to fertilise their own eggs. Self-fertilisation does occur in some species, particularly where the likelihood of successful cross-fertilisation is very low. In the majority of hermaphrodites, however, there are mechanisms to avoid this possibility. Inbreeding, as opposed to fertilising the eggs or ovules of a non-relative, may be harmful because it reduces reproductive success. This arises immediately where any harmful recessive genes are expressed in the homozygous state in the offspring or, in the long term, through a reduction in genetic variation.

In some examples it is possible to analyse the way in which resources are allocated between male and female function. Generally, such resources would be expected to maximise an individual's reproductive success, but it is not always clear how this is achieved. The formal problem of sex allocation is, however, similar to that of optimal sex ratio (Section 4.3) which describes how parents should allocate energy between sons and daughters.

Hermaphroditism is probably not a primitive condition but has evolved a number of times in different taxa in response to particular conditions. Because hermaphroditism permits an individual to function as both male and female it could potentially achieve greater reproductive success than it would as a pure member of either sex. The exact benefit would depend on the relationship between investment and reproductive success. For example, if only a small part of the energy invested in male function produced a high proportion of male success then a prudent individual might maximise its overall success by becoming hermaphrodite and investing in female function as well. Often, however, it is found that hermaphrodite species have close relatives that are dioecious because there are also significant costs and constraints that prevent hermaphroditism from becoming more widespread.

3.2 Sex change

Sex change may be either from male to female, described as ***protandrous*** or 'male-first', or from female or male, described as ***protogynous***. Two principal conditions appear to favour sex change. First, individuals change sex as they grow larger because of size-related reproductive expectations. Among plants and invertebrates sex change from male to female may occur when female fecundity is strongly size-related but male reproductive success increases only marginally with size. It also seems to be generally true that animals which change sex are those which have indeterminate growth patterns. Second, sex change occurs because of a temporally patchy environment which favours an individual functioning as one sex only in particular time periods.

Constraints on sex change are usually physiological or metabolic. One problem is that reproduction by one sex, generally female, may be relatively expensive so growth is sex-specific. Selection should then delay to larger sized individuals the reproductive function of the sex which pays the growth cost. In other words, there is unlikely to be enough energy available to reproduce successfully as a small female and also grow into a large size class. In addition there are the metabolic costs of sex change compared to the reproductive costs of staying the same sex. Small male mammals may have low reproductive success but the costs of changing sex for a mammal are prohibitively high. The time taken to make the sex change will also reduce the time available for reproduction. An individual may not live long enough to make sex change worthwhile despite possible size advantages.

3.2.1 Sex change in algae and vascular plants

Many of the seaweeds are simultaneous hermaphrodites but *Cytoseira* shows a distinct age-related pattern of sexual function. The youngest plants go through an antheridial period and function solely as males. Subsequently, as the plant gets larger, the older parts change to produce oogonial conceptacles while antheridial conceptacles are still produced at the growing tip. At this point the plant as a whole is functioning as a simultaneous hermaphrodite but there is a defined age and size-related separation in function within an individual.

Other plants go through a complete change in reproductive function from season to season. This is sometimes described as labile (as opposed to developmental) sex change. Jack-in-the-Pulpit, *Arisaema triphyllum*, is a long-lived herb of deciduous forests in north America. It overwinters as a corm and thus has to rebuild the above-ground part of the plant every year. Plant development is determined by the leaf buds that emerge in late May and they in turn depend on the amount of nutrients stored in the corm the previous year. In studies of several populations in the United States the flowering pattern of a large number of plants has been followed over several years. Individuals were identified by adjacent numbered stakes. Inflorescences on the largest plants were exclusively female within a flowering season while those in intermediate categories were male and small plants did not flower.

Between seasons plants changed in size according to their growth and hence storage input from the previous year. As plants changed in size so they also changed sex (Table 3.1). Thus, small female plants were likely to change to male plants while large males, if growing, were likely to become female the next season. Sex change in this case is facultative rather than protandric because, although smaller individuals are male and reproductive success is size-related, size is not age-related. How should reproductive strategy change with size for plants functioning as male or female? The sex ratio (see Chapter 4) should become female biased in the size class in which female reproductive success exceeds that of a male.

Table 3.1 Sex change in *Arisaema*, Jack-in-the-Pulpit. Individual plants change in sex between seasons and larger plants are usually female. Reproductive success increases more strongly with size for females than it does for males. J.L. Doust and P.B. Cavers (1982, *Ecology*, **63**, 797–808) recorded the proportion of plants which changed between several reproductive states or were absent between years.

		Condition in 1980			
		Female	Male	Vegetative	Absent
Condition in 1979	Female	0.72	0.0	0.28	0.0
	Male	0.59	0.12	0.29	0.0
	Vegetative	0.28	0.14	0.57	0.01
	Absent	0.0	0.0	0.62	0.38

Quite a number of plants, such as *Impatiens biflora* which grows alongside *Arisaema*, have hermaphroditic flowers which are strongly protandric. Male and female function are so temporally displaced that selfing is improbable. However, this is not quite the same as sequential sex change since the energetic investment to male and female function probably occurs at about the same time. There are rather few well documented cases of other plants with sex change. As with animals, the known examples are taxonomically widespread and have relatives which do not change sex.

Generally, sex change to male is associated with stress such as cold winters or drought but artificially heavy pruning can induce a change to femaleness.

3.2.2 Sex change in animals

Sex change has been described in many taxa including molluscs, polychaetes, arthropods (but not insects) and bony fishes. It seems generally to be true that the majority of invertebrate sex changers are protandrous. Smaller or younger males can function as effectively as larger ones, while big females are significantly fitter than small females. Protandrous sex change has been widely recorded among prosobranch molluscs including the slipper limpet, *Crepidula fornicata*. *Crepidula* are long-lived molluscs which also brood their young. They are found in muddy bays where they aggregate in stacks of up to 20 individuals. Stack structure is characteristic and the largest animal at the bottom is always female while small individuals on top are male. Smaller animals are relatively mobile and may be able to travel between stacks, so having potential access to many females. Larger individuals are immobile but since fecundity is size-related they produce many eggs. Sex reversal appears to be controlled by social influences. Males do not grow in the presence of females, but the removal of a female causes males to grow faster and to change sex.

In non-gregarious species there is no evidence of socially influenced sex change, but it may be affected by population dynamics. In pioneer populations of sessile bivalves, such as oysters, *Crassostrea*, where spats settle on artificially clean substrate the sex ratio in the one-year-old size class is about 0.5. In age-structured populations containing larger and older individuals most of the one-year-olds are males. A similar pattern may be seen in the common limpet, *Patella* (Fig. 3.1).

Pandalus jordani is an hermaphrodite shrimp which, like all known sex-changing Crustacea, is protandrous. Populations of *Pandalus* have large annual fluctuations in age and size structure and individuals respond to such changes by altering the age at which they change sex. Because larger female shrimps are of greater economic value, and many of the commercially exploited Crustacea are also protandrous hermaphrodites, the impact of fishing on populations has been to lower the age (or size) of sex change. This flexible response to sex change would be predicted if each individual is to maximise its reproductive success but it is not clear what cues the animals use to determine their local response.

In some populations there are two distinct life-history patterns. Some animals change sex while others exist as females throughout their lives. An individual should mature immediately as a female where its reproductive success as a pure female is relatively large compared to an hermaphrodite. Examining the distribution of early maturing female *Pandalus borealis* on a geographical scale a close relationship emerges between data on population egg production and sex change and the theoretical predictions. Northern populations have slower growth rates and longer life expectancy

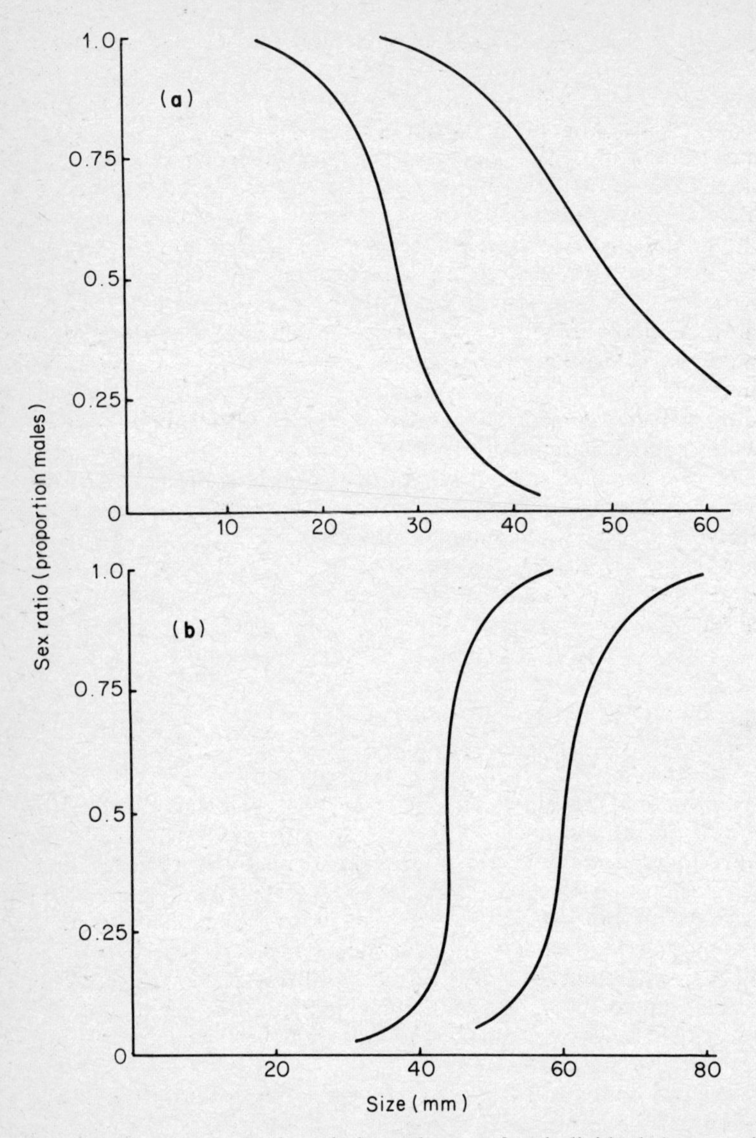

Fig. 3.1 In some animals and plants the sex of an individual changes with size or age. Among plants, sex may change back and forth several times but in animals there is usually only one sex-change event. The stage at which this happens in an individual's life is not fixed but may depend on local ecological conditions. In these two cases data have been drawn from two populations of each species to illustrate the variation between sites: (**a**) protandry in *Patella vulgata*, a marine limpet, at Batten Bay near Plymouth, England may depend on both size and age; (**b**) protogyny in *Anthias squamipinnis*, a planktivorous reef fish, at Aldabra atoll in the Indian Ocean depends on size and social factors. (Adapted from Figs 12.3 and 12.1 in Charnov, E. (1982). *The Theory of Sex Allocation*. Princeton University Press, New Jersey.)

and at higher latitudes the average length of time spent in the male phase extends from less than one to over two years.

Unlike most invertebrate examples sex change in fish is usually protogynous, from small female to big male (Fig. 3.1). Very little is known of the protandrous groups. Sex change has been studied in detail for the coral reef parrot fishes (Scaridae) and wrasses (Labridae). Typically, females are smaller and often duller than males but there are also cases where small fish may be male if there are opportunities for them to breed. Among wrasse, *Thalassoma* males establish large territories at reef edges which are favoured by females who come there to spawn about once a day. There are a limited number of such territories but small, non-territorial males may have mating opportunities if they join group-spawning encounters.

The cleaner-fish, *Labroides dimidiatus*, is also territorial and male territory owners control harems of 5–6 mature females. There are no small male *Labroides* because females remain and feed at stations within the male's control. The male is the largest individual but the females also form a size-based hierarchy. If the male dies the oldest, dominant female takes over the territory and within hours of the male's death is controlling the harem and behaving as a male. In less than two weeks this individual can produce sperm.

3.3 Simultaneous hermaphroditism

One advantage of functioning as both a male and female simultaneously is that any two individuals encountering one another are able to mate. This is likely to be advantageous among species that are relatively immobile or at very low density. In both cases encounter rates are comparatively low but opportunities for mating would be doubled by hermaphroditism. A second condition favouring hermaphroditism is where the costs of or investment in male and female function can be spread out over some period. This has two advantages in that it avoids competition between the male and female function for available resources, and it permits a greater total reproductive investment.

There are two major constraints. First, it may not be morphologically possible to carry reproductive apparatus for both sexes. This would appear to be true for many vertebrates but the metabolic costs of simultaneously producing two sets of gonads and gametes may be a wider restriction. Second, there is a risk of self-fertilisation which has incorrectly been assumed to be an advantage to hermaphrodites. In some organisms there are elaborate structures to prevent such a possibility. However, where self-fertilisation occurs there is often a reduced investment in male function since fewer male gametes are required to ensure reproductive success. A third restriction that may apply to some animals is that successful function as a member of either sex may be a product of behaviour or experience as a member of that sex. Even where young males are not breeding they may be gaining experience in combat or courtship that would be lost or inappropriate if they also functioned as a female at that stage.

3.3.1 Simultaneous hermaphroditism in vascular plants and algae

Less than 3% of the British flora are dioecious and the vast majority function as simultaneous hermaphrodites. The distribution of function may be between unisexual flowers on one plant (monoecy) or within a flower (monomorphism). The arrangement of sexes within a monoecious plant is presumably designed to maximise the effectiveness of each sex's function. Among wind-pollinated conifers the female cones are usually at the top of the tree and the male cones are below. This is believed to help in three ways. First, it ensures that seeds are produced by outcrossing because male and female are spatially separated. Second, pollen drifts down from the canopy into the open air column rather than into the tree. Third, pollen is received from a variety of other individuals which are unlikely to be close neighbours. Other plants alter their floral sex ratio according to local conditions, so *Quercus gambelii* tends to be more female when growing under a canopy where pollen receipt is high but pollen donation is unlikely to be successful.

In grasses, which are also wind-pollinated, the pollen is produced at the top of the stem. This promotes more efficient pollen dispersal in dense plant communities close to the ground. The 'males-above' arrangement in *Delphinium*, is, however, explicable in terms of pollinator behaviour which affects both the efficiency of pollen dispersal and outcrossing. Larkspurs (*Delphinium* species) have vertical monoecious inflorescences which contain female and male flowers. They are pollinated by bumblebees, *Bombus*, which characteristically start foraging at the bottom of an inflorescence and gradually move vertically up among the flowers (Fig. 3.2). The nectar abundance decreases with flower height and the bee moves on to the next plant before reaching the top. Significantly, female flowers are restricted to the bottom of the inflorescence while staminate flowers occur above them. So, as the bees move between plants they pick up pollen loads just before leaving and moving to the most receptive part of another plant. This ensures outcrossing but since the bees tend to fly short distances between feeding the pollen will be transferred to plants with similar local adaptations.

Flowers are elaborate and costly structures designed to attract the attention of pollinators: a function of benefit to both male and female. A possible advantage to monomorphism is that the cost of the flower is spread across male and female functions rather than being borne separately by both, but there may then be a requirement to avoid self-fertilisation. An example of polymorphic flower types in a monomorphic plant also encourages outbreeding. Primrose, *Primula vulgaris*, has two morphs, one of which has a long style and short stamens (the pin type) and the other has the reverse arrangement (the thrum type). This is called heterostyly.

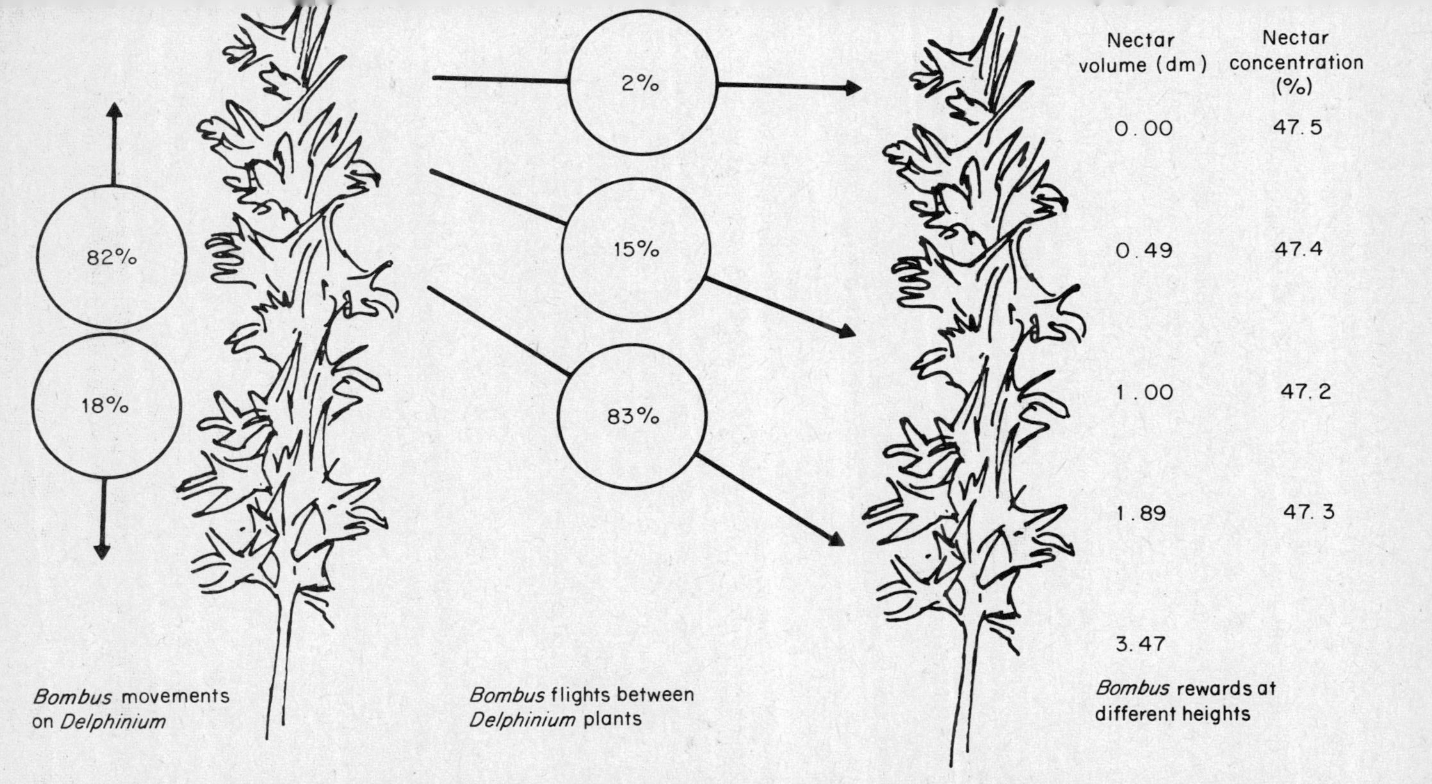

Fig. 3.2 Flowering spikes of the larkspur *Delphinium* have a 'males above' flower sequence. This is explicable in terms of the foraging behaviour of their main pollinator, *Bombus* bumblebees. Pollen is collected by the bee at the top of one spike and is carried directly to the bottom of the next spike. (Pyke, G.H. (1978). *Oecologia*, **36**, 281–93.)

Primroses are pollinated by a variety of insects which differ in behaviour. Flies, for example, tend only to visit the top part of the flower and so transfer pollen from thrum anthers to pin styles. This ensures that the flowers are regularly outcrossed, but self-compatible homostyle variants of *Primula* have been recognised. In some populations homostyly includes up to 25% of the plants. In one population in Somerset, England, there is a very high rate of selfing among homostyles which is combined with a significantly greater seed set than among the heterostyle morphs. This is accentuated when the weather reduces the abundance of pollinators. Since high seed set is advantageous the homostyle variant should invade other populations and replace heterostyly. There are two possible reasons why it does not. First, selfed homostyles may suffer long-term reductions in viability. Second, homostylous plants are also protogynous (the stigma is receptive before the anthers dehisce) and thus may often be fertilised by outcrossing in normal years.

On the seashore there is a well-marked zonation of brown seaweeds or wracks (Fucales) across the intertidal zone (Fig. 3.3). At the top of the shore *Pelvetia canaliculata* and *Fucus spiralis* are hermaphrodites. Near mean sea level *Fucus vesiculosus*, *F. serratus* and *Ascophyllum nodosum* are dioecious. This pattern may be associated with the lower probability of cross-fertilisation on the upper shore. The gametes of seaweeds are developed on both surfaces of the frond in tiny conceptacles. Usually these are then released into the sea for external fertilisation. Because the upper shore is covered for a shorter part of the tidal cycle it is likely that gamete mobility will be very poor. In the hermaphrodites the antheridia and oogonia are borne within the same conceptacle, producing a high probability of self-fertilisation. Inbreeding may be a means of ensuring that a high proportion of eggs will be fertilised and in *Pelvetia* viable young plants have been observed within the conceptacle. Comparing reproductive investment in *Fucus*, the dioecious species which outcross produce approximately ten times as much sperm as *F. spiralis*. Unfortunately, restrictions on gamete mobility are unlikely to be a complete explanation since sea-oak *Halidrys*, which is also hermaphrodite, occurs low on the shore in pools near the dioecious *F. serratus*.

3.3.2 Simultaneous hermaphroditism in animals

The distribution of hermaphroditism through the animal kingdom appears arbitrary. Among the Platyhelminthes the triclads are usually hermaphrodite but also reproduce by fission while rhabdocoels may be hermaphrodites or sex changers and trematodes are often dioecious. Triclads have two complete sets of reproductive organs and also have well-developed structures to avoid self-fertilisation. This additionally gives them the opportunity to digest the sperm of copulatory partners: a useful energetic bonus. Hermaphrodite gastropods offset some of the costs of sex organs by having a common ovo-testis, but this severely increases the risks of self-fertilisation and gamete production is sequential rather than

Fig. 3.3 Because of contrasting ecological conditions the favoured mode of reproduction among fucoid seaweeds changes along a downshore transect. At the top of the tide, where conditions are most extreme, there are hermaphrodite (HA) species, but those species which are lower down the shore and consequently get covered by all tides tend to be dioecious (DO). From the top the species illustrated are: *Pelvetia canaliculata*; *Fucus spiralis*; *Fucus vesiculosus*; *Ascophyllum nodosum*; *Fucus serratus*; *Himanthalia lorea*. (Vernet, P. and Harper, J.L. (1980). *Biological Journal of the Linnean Society*, **13**, 129–38.)

simultaneous as in triclads. Among the molluscs there are both dioecious (gonochoric) and hermaphrodite groups. It may be true that marine species are more frequently dioecious. All of the relatively mobile Cephalopoda have separate sexes as do most marine Bivalvia though some are also sex-changers. Many species of freshwater bivalves have hermaphrodite populations. Along the Rhine the freshwater mussel, *Anodonta*, occurs both in the main river and in nearby oxbow ponds which have been cut off from the Rhine for various lengths of time, documented up to 300 years. The riverine parent population is exclusively dioecious but recently isolated populations show sex ratios heavily biased towards females and in the oldest isolates the animals are exclusively hermaphrodite.

Another example of an animal which is hermaphroditic at low densities is the rice-paddy shrimp *Triops cancriformis*. *Triops* is found from north Africa, where it is common and exclusively dioecious, up through Europe, where it becomes increasingly rare. Northern populations appear to be spreading and there, on the edge of its range, *Triops* is hermaphrodite. Similarly, new populations colonising rice paddies in Florida are also hermaphrodite so it seems unlikely that it is the northern climate which causes the change in sexual strategy.

Low density populations and reduced mobility are often a characteristic of metazoan parasite populations. Requirements for access to mating partners may lead to environmental sex determination (Section 2.3) or, commonly, to hermaphroditism. Among the Annelida, marine polychaetes are predominantly dioecious while leeches are invariably simultaneous hermaphrodites. Many freshwater leeches, however, are non-parasitic, highly mobile and occur at substantial densities. *Glossiphonia* is a common genus in British lakes and rivers which broods its young. It lays cocoons which it retains on its belly, caring for the young for an extended period. In these species, male expenditure occurs in the early period up to copulation while female expenditure occurs after fertilisation and the functions are thus separated in time. A related species, *Helobdella stagnalis*, has well marked periods of sperm production alternating with cycles of egg production so energy expenditure is maintained over a prolonged period.

Like the triclads, leeches also have sperm digesting organs. Mating occurs as much to donate sperm as to have eggs fertilised. There is thus a conflict of interest in cross-copulation because an individual seeks to conserve its eggs, in order to copulate again, while fertilising as many other eggs as possible. The presence of sperm-digesting tissue leads to the evolution of mechanisms to avoid the partner's attempts to misuse sperm donations. Leeches do this by injecting their motile sperm directly into the partner's hemocoel and, since there is also internal tissue to control sperm movement, they grasp one another in copula to ensure that the injection is placed in an area underlain by the appropriate vector tissue.

Among the vertebrates only a small number of fish are simultaneous hermaphrodites. Fish usually fertilise eggs externally and in the case of a coral reef hermaphrodite, the black hamlet *Hypoplectrus nigricans*, this

Table 3.2 Egg trading in the Black Hamlet, *Hypoplectrus nigricans*. Each fish is hermaphrodite but will only trade eggs to be fertilised if a spawning partner reciprocates. E.A. Fischer (1980, *Animal Behaviour*, **28**, 620–33) showed that spawning bouts without reciprocation tended to end more rapidly than those where both partners provided eggs.

	Number of spawns as female					
	0.5–1	1.5–2	2.5–3	3.5–4	4.5–5	5.5–9
No reciprocation	14	6	2	1	1	0
With reciprocal egg deposition	15	21	19	29	34	32

has led to a complex system of egg-trading behaviour. The hamlet spawns in pairs in which only one functions as a male or a female at a time. While one partner, usually the one initiating a spawning bout, releases eggs the other fertilises them. This ensures that self-fertilisation should never occur. However, not all eggs are released in a single parcel. In the next part of a sequence the partners reverse roles and the fish that first released eggs now acts as a male while the other partner reciprocates by providing eggs. In a field study, 20 out of 24 bouts where no reciprocation occurred ended after the initiator had spawned twice as a female. By contrast, in 150 bouts where reciprocation took place up to nine and on average at least four spawning events were recorded (Table 3.2). So the black hamlet ensures reciprocation by carefully parcelling its eggs out in a trading sequence. This trading may account for the maintenance of hermaphroditism, since pure males would have no eggs to trade, but there does not appear to be any reason why it evolved in the first place.

4 Sex Ratios

The sex ratio of a population of animals or plants is a measure of the relative proportions of males and females. It is a ratio that can be expressed in a number of different ways. Sometimes it may be useful to quote the absolute number of males and females in a sample. For example, in England and Wales in 1973 there were 1 917 000 males and 1 816 100 females up to four years of age. Alternatively the number of one sex may be quoted relative to a fixed number of the other. So for every 100 females up to four years of age in 1973 there were 105.6 males. This type of reference system may be used in another way to express the number of members of one sex relative to a single individual of the other sex. A proportionality of 1:3 indicates a ratio of three females to every one male. This could also be described as having one male in every four individuals, or a sex ratio of 0.25 (25% male) or 0.75 (75% female). The diversity of sex ratio conventions can lead to confusion and it is important to be clear which system is being used. In this book the convention that a population with 25% males has a sex ratio of 0.25 will be used.

4.1 Sex ratio variation

Sex ratios frequently vary with age. The ***primary*** sex ratio is the sex ratio at conception, immediately after gametes have fused to form a zygote. Variation in the primary sex ratio may be a product either of the balance of gamete production from the heterogametic sex or of differential mortality or success rate among the heterogametes. In certain cases the primary sex ratio may be subject to manipulation by the mother, for instance in haplodiploid species where the mother can select which eggs are fertilised (Section 2.2).

The ***secondary*** sex ratio is that at the end of parental investment. Marine fish, such as the cod-fishes (Gadiformes), release vast numbers of gametes directly into the sea with no subsequent parental care. Amongst higher vertebrates, where there is post-natal investment, the primary and secondary sex ratios may differ due either to pre-natal differential viability among embryos or to differential mortality during post-natal care.

The ***tertiary*** sex ratio is the balance of adult individuals in the population as a whole. This will obviously differ from the secondary sex ratio where there are sex-related differences in ecology and longevity. The ratio of sexually active males to females available for fertilisation may also vary at any particular time or place. We may then be able to calculate an ***operational*** sex ratio.

The difference between the tertiary and operational sex ratios is illustrated by the Common Toad *Bufo bufo*. Toads migrate each spring to traditional ponds and lakes to breed. During the breeding season the tertiary sex ratio is male biased. Possible reasons for this bias include greater female mortality or earlier maturity of male toads, which are the smaller sex. The operational sex ratio is even more skewed in favour of males because they remain at the breeding site for as long as there are females returning to spawn. By contrast, females stay only a few days and leave after egg-laying. Since females do not all arrive at the same time, most of the males will be competing for only some of the females at any one time.

Among dioecious plants such as the lily, *Chamaelirium luteum*, the operational sex ratio will depend on the flowering patterns of males and females. In one population only about 10% of the plants flowered at any one time and there were always more males in flower than females. This was not due to any sex bias among seedlings (secondary sex ratio) but because males tended to flower more frequently.

4.2 Analysing sex ratios

As well as the principal sex ratios described above it may also be useful to split males and females into discrete age or size classes. There is a very well marked pattern of age-related changes of sex ratio in humans (Fig. 4.1). The sex ratio at birth is usually male biased: in 1980 there were a total of 380 000 male births and 360 000 female births in the United Kingdom. This is by no means an accurate estimate of the primary sex ratio, since it fails to take into account any differential loss between the sexes at very early stages of pregnancy. It is generally believed that the sex ratio at conception is more male-biased than at birth.

Secondary sex ratio in humans is a matter for debate because it is difficult to determine the point at which parental investment stops, if ever. However, the death rate among boys up to the age of 15 is about 15.2 per thousand and among girls is about 12.0 per thousand. This still leaves an excess of males in the population which persists up to the late thirties. In the adult breeding population, which we arbitrarily take to include the age classes 16–44 years old, the tertiary sex ratio in 1980 was about 0.53. Male mortality continues to be higher and, by the age of fifty, females begin to outnumber males. In the 75+ age group the sex ratio has fallen to 0.3.

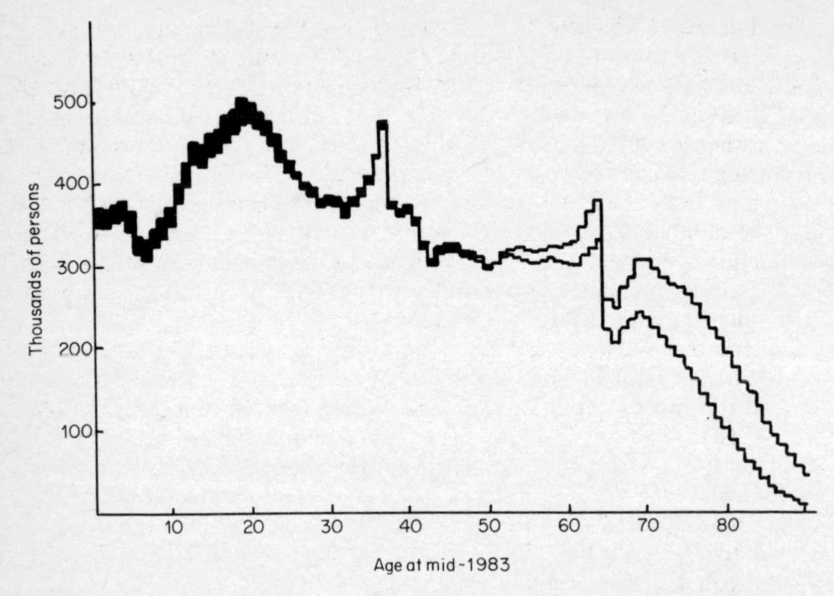

Fig. 4.1 There is a characteristic pattern of changing sex ratio across age classes in human populations. This figure illustrates the structure of the British population, according to age and sex, in 1983. Despite year to year variation in the number of births, males consistently outnumber females in age classes up to the late forties. Solid shading between the lines indicates an excess of males while open blocks indicate an excess of females. (Data from the Office of Population Censuses and Surveys.)

4.3 Evolution of the sex ratio

Most secondary sex ratios are close to 0.5. This may seem surprising since, in many plants and animals, a male can fertilise many females. So why don't species produce only as many males as are necessary to ensure that most females can find a mate?

One suggestion is that an equal sex ratio is an inevitable consequence of the pattern of segregation of the sex chromosomes at meiosis (Section 2.1). Thus, for mammals, half the sperm will carry the male-determining Y-chromosome so that half the fertilised eggs will be male. This argument is causal rather than functional and explains what happens rather than why it happens. It presupposes that the sex ratio is not subject to the forces of natural selection. The evolutionary reason, now generally accepted, for a secondary sex ratio of 0.5 was first put forward by R. A. Fisher in 1930 and is known as Fisher's sex ratio theory (Fisher, 1958).

The theory can be illustrated by the following hypothetical example. Imagine a population with an excess of females, where a male can obtain several mates. Each male in these circumstances will then have, on average, a greater number of offspring than each female. Each female produces about the same number of offspring, but a female would have a greater number of grandchildren than other individuals if she produces sons rather than daughters. If this trait is heritable then the male-producing tendency will spread in the population. If male births increase to such an extent that the sex ratio exceeds 0.5 it then becomes better to produce daughters. Put simply: natural selection will act to favour the production of the rarer sex. The outcome is an equilibrium sex ratio of 1:1, which is also an evolutionarily stable strategy (ESS) since it can be bettered by no alternative strategy.

Fisher's theory has a number of important applications and assumptions. First, it implies that we should expect to find a 0.5 secondary sex ratio at the end of parental care in both monogamous and polygamous species (see Chapter 6). In some species, for example the Northern elephant seal *Mirounga angustirostris*, one male may mate with many females while other males may not mate at all. Despite the greater variation in mating success between males than between females the average pay-off to each sex is the same. Thus the potential reproductive success of a son is equal to that of a daughter.

Secondly, mortality after the period of parental care does not affect the secondary sex ratio. This is certainly true in the case of Richardson's ground squirrel *Spermophilus richardsoni* where the secondary sex ratio is 0.5 but male mortality is subsequently much higher than female (Section 4.4).

Thirdly, the argument implies a 0.5 sex ratio only where the cost of producing a male and a female are the same. If one sex is more expensive to produce then the secondary sex ratio should be biased in favour of the cheaper sex. In other words, Fisher's theory can be restated in the form: invest equally in both sexes. Data on Hymenoptera suggest that this does happen. Asocial bees and wasps go through larval stages in small holes dug by the mother. In the adult stages males and females differ in size and, since all growth occurs in the natal cell, this size must depend on the amount of food provided by the mother. The smaller sex will require less food and should be cheaper to produce. Among species where females are the larger sex the majority have male biased sex ratios (Table 4.1, see also Section 2.2).

4.3.1 Local mate competition

Fisher's sex ratio theory assumes that individuals mate at random in the population. In many cases such 'panmictic' mating is unlikely. Among those plants whose seeds are not dispersed widely seedlings are likely to grow near their parent and in close association with their siblings. In these cases significant levels of competition between brothers and sisters for mating or breeding opportunities should be expected.

In some species there are examples of evidently non-random mating that

Table 4.1 Sex ratio bias in Hymenoptera. Among solitary bees and wasps females are usually larger than males. The sex of offspring can be selected by the mother when she lays her eggs in hosts. Since males are smaller they should be cheaper to produce and therefore should be produced in greater quantities. Data gathered by R.L. Trivers and H. Hare (1976, *Science*, **191**, 249–63) illustrate this bias.

	Weight (g)			
Species	Males	Females	Weight ratio	Sex ratio
Bees				
Agopostemon nasutus	6.4	9.7	1.52	0.68
Anthophora abrupta	36.7	58.0	1.58	0.62
Chilicola ashmeadi	0.5	0.8	1.6	0.74
Euplusia surinamensis	123.6	148.7	1.20	0.59
Hoplitis anthocopoides	12.3	11.6	0.94	0.66
Nomia melanderi	31.9	25.5	0.80	0.50
Osmia excavata	16.3	24.9	1.53	0.63
Wasps				
Antodynerus flavescens	14.6	22.2	1.52	0.61
Chalybion bengalense	8.7	19.6	2.25	0.60
Ectemnius paucimaculatus	3.0	4.0	1.33	0.65
Pasaloecus eremita	1.6	3.4	2.13	0.41

are accompanied by sex ratios consistently divergent from 0.5. The reason for these extraordinary sex ratios was explained by W. D. Hamilton in terms of the restrictions on mating opportunities. The mites, *Adactylidium*, mate exclusively with siblings: a female is always inseminated by her brother. In fact, the male role in this mite is reduced to a transitory life stage serving only as a sperm producer. The immature stages develop entirely within the mother's body where they feed on her tissues until reaching maturity. Only a single male is produced amongst a brood of six to nine; sex ratio is controlled by the mother and males are produced by arrhenotoky, from unfertilised eggs (Section 2.2). The male mates with his sisters within the mother's body and he dies before he is born! A mother will maximise her fitness by producing as few sons as are required to ensure that every daughter's egg production is completely fertilised. In *Adactylidium* this is satisfactorily achieved by ensuring that one son is produced per brood. Hamilton recorded about 25 examples of insects and mites with strongly biased progeny sex ratios, usually with a single son per batch. In all cases sib-mating occurred before dispersal from the larval site (Table 4.2).

Because Hamilton's examples referred to cases where mating between siblings was obligatory feature of the system, it has been incorrectly assumed that inbreeding is the only factor in the evolution of biased sex ratios. This is not so. The other important factor is the extent of competition between same-sex siblings, often brothers, for mating opportunities.

Table 4.2 Local mate competition in parasitoid arthropods. Some parasitoids mate with their siblings before leaving their host. Under these circumstances brothers compete with one another for mating opportunities. W.D. Hamilton (1967, *Science*, **156**, 477–88) suggested that it would be adaptive for mothers to invest more in daughters than sons; we should then expect a female biased sex ratio among emerging progeny.

		Number of progeny	
Parasitoid species	Host	Male	Female
Nasonia vitripennis	Fly pupa	2	19
Prestwichia aquatica	Insect egg	1	8
Anaphoidea nitens	Weevil ootheca	1	3
Xyleborus compactus	Twig tissue	1	9
Limothrips denticornis	Grass plant	3	20
Blasophaga penes	Wild fig	22	235
Acarophenax tribolii	Mother	1	14

In the case of *Adactylidium* it is obvious that if several sons were produced they would be competing only with one another to mate. The mother's chance of grandchildren is not improved by producing more sons: indeed, male production at the expense of females is likely to decrease the number of her grandchildren. Overall, because reproductive output is often limited by the availability of eggs rather than sperm, local mate competition would be expected to occur more frequently between related males than related females.

4.3.2 Local resource competition

Another form of competition can also affect the equilibrium sex ratio. It involves competition not for mates but for other resources. It is similar to local mate competition, however, in that it concerns competition between related members of one sex. If, for example, related females but not males compete for resources, such as food or nesting sites and this competition has an effect on fitness then it would be expected that secondary sex ratios would be biased in favour of males.

Such a situation appears to occur in the Thick-tailed Bushbaby, *Galago crassicaudatus*. The galagos live in loosely-bonded groups in the East African bush. Young females settle close to their birth site whereas sons disperse. These animals are omnivorous, feeding on gum from *Acacia*, insects and fruits but, despite this diverse diet, their food supply is often localised in abundance. These resources are also renewed fairly rapidly but can be depleted in minutes and are only rarely found in sufficient quantity to support pregnant or lactating mothers. The sex ratio in the wild is usually male biased and this finding is supported by museum and zoo records (Table 4.3).

Even when mature, daughters tend to associate with their mothers and

Table 4.3 Local resource competition in *Galago crassicaudatus* (see text). Field data and skin collections indicate that population sex ratios are biased towards the more wide-ranging males (A.B. Clark (1978). *Science*, **201**, 163–5).

Data source	Sex of specimen			Apparent sex ratio
	Male	Female	Unknown	
Museum skin collections of 12 subspecies	483	273	182	0.62**
Reported captive births	209	141	100	0.60**
Births in captivity over five year book records	126	96	65	0.57*
Field study births	13	4	8	0.76*
Field study totals	21	9	4	0.70*

* = $P < 0.05$ ** = $P < 0.005$

engage in mutual grooming as well as sharing sleeping perches. Their home ranges overlap and as a consequence they use the same food gathering sites. Since females carrying young are restricted in mobility they remain in areas of their range where food supplies are more abundant and this seems to be essential to their reproductive success. If mothers and daughters or sisters produced young at the same time they would inevitably compete for the most favourable sites and their reproductive success would probably decline. This competition becomes worse as the number of females present in the area increases. Sons, on the other hand, range widely and have no effect on the reproductive success of their mother, their sisters or their brothers. Since a daughter's fitness is reduced by competition with her sisters but a son's fitness is unaffected by the sex ratio, then the mother's reproductive success will be increased by producing relatively fewer female progeny.

4.3.3 African hunting dogs

The African Wild Dog, *Lycaon pictus*, provides an interesting example where two different sex-related interactions between siblings may be affecting the sex ratio. Here, the concept of cooperation between generations as an influence on offspring costs can be introduced. If sons or daughters remain with the parents and subsequently assist in the rearing of a further litter they are effectively repaying the cost to the parents of raising them. Parents which raise more of the assisting sex can invest less in future litters when being assisted. This is functionally similar to having sex differences in the initial cost of raising that sex so there should be selection for parents to produce more of the assisting (cheaper) sex.

Lycaon packs have a well defined social structure with one dominant breeding pair and a number of related males plus recent offspring. In an area of Tanzania, 52 of the packs had only a single female and the average was 2.1 females while the mean number of males was 4.1. Sisters disperse from packs together but brothers remain at the natal site. The subordinate males in a pack, either sons or brothers, assist in cooperative hunting and help to feed the pups. They thereby contribute to the rearing of their young relatives and are effectively the cheaper sex to raise. Sisters, on the other hand, are dispersing together and subsequently compete for access to male groups in order to become the sole breeding female in the unit. The sisters may thus be engaging in local mate competition.

So, whether considering the sisters or the brothers, a secondary sex ratio biased towards males would be expected. In fact, the sex ratio of one-

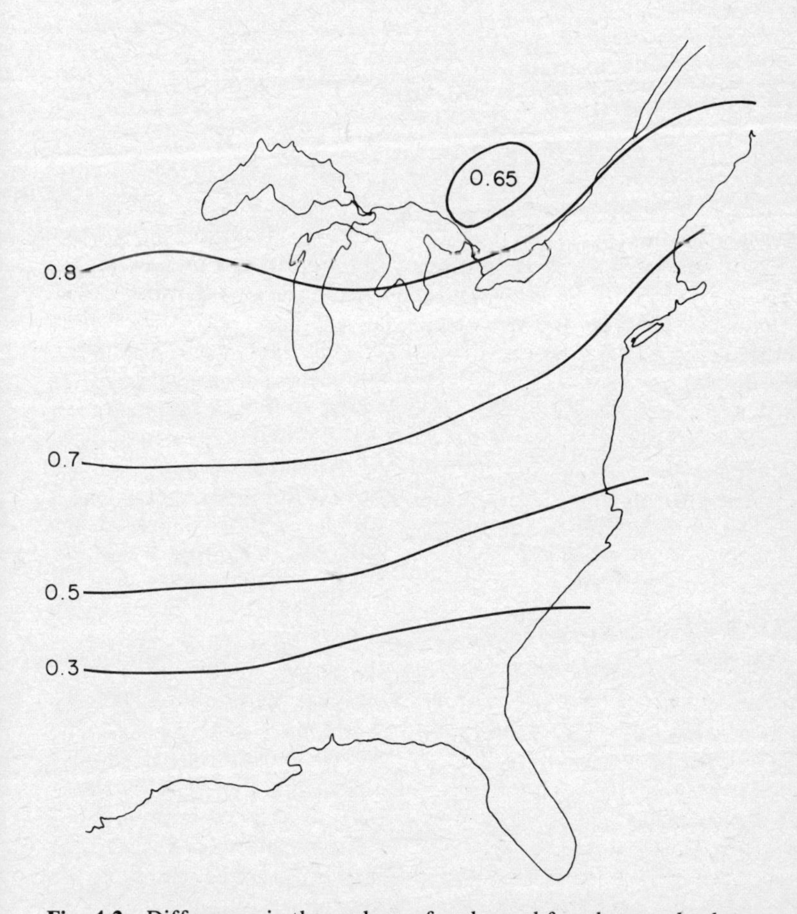

Fig. 4.2 Differences in the ecology of males and females may lead to census results that suggest odd sex ratio patterns. This figure describes the distribution and sex ratio (proportion males) of wintering dark-eyed juncos *Junco hyemalis* in the eastern United States. Sex ratios change with latitude because males overwinter further north than females. (Ketterson, E.D. and Nolan, V. (1976). *Ecology*, **57**, 679–93.)

month-old pups was 0.59 in a sample of 96. The mechanism by which this bias is achieved has not yet been identified.

4.4 Sex ratios in nature

Many other factors will affect the apparent sex ratio. Because of differences in the behaviour, ecology and thus catchability of males and females it can be surprisingly difficult to obtain a sample that allows us to make an accurate estimate of the sex ratio of the population. Among mammals, males are often more active or wide ranging than females and are thus more likely to be trapped. The home range of a male weasel, *Mustela nivalis*, is about two to three times larger than for a female and, at low trap densities, the sex ratio is male biased, often around 0.75 (see Fig. 4.3b). If trap spacing is altered then the sex ratio also changes and at very high trap densities the ratio appears to become female biased.

Museum collections of skins of the Dark-eyed Junco, *Junco hyemalis*, produce a pattern with a 0.3 sex ratio from the southern United States and 0.7 from the northern States. A similar pattern appears in the detailed analyses of the American Audubon Society's annual Christmas bird count (Fig. 4.2). The fact that the counts are taken at Christmas is the critical factor: males overwinter farther north than the more migratory female juncos. Males are larger and are better adapted to face the harsher winter climate but it is not clear what advantages they gain from remaining in the north. Other birds, including the Chaffinch, *Fringilla coelebs*, in Europe and the Red-billed Quelea, *Quelea, quelea*, in Ethiopia, also show seasonal patterns of local sex-ratio disparity associated with breeding cycles.

For plants, sex ratio differences in habitat and geographical distribution may be more local. There are often ecological differences among dioecious plants and it is often the case that male (staminate) plants are found in harsher, drier conditions than female (pistillate) plants of the same species (Section 5.3). There is also a general, unexplained trend among dioecious plants for perennials to have male-dominated populations and for annuals to have female-dominated populations. In the tropics about one third of dioecious forest tree populations are male biased, the rest having equal numbers of the two sexes. Females, on the other hand, are generally more numerous in populations of temperate herbs such as Dock, *Rumex* (Fig. 4.3a), and Campion, *Silene*, possibly due to an excess of vegetative over sexual reproduction.

Finally, differential predation can also have a major effect on the sex ratio. Richardson's Ground Squirrels, *Spermophilus richardsonii* (Fig. 4.3c), have tertiary sex ratios around 0.25 despite secondary sex ratios close to 0.5. This appears to be due to the different activity patterns of the sexes. In Alberta, male squirrels emerge from hibernation earlier in the spring when they are likely to be more at risk to predation by owls and migrating hawks (*Buteo* species). Males also range over a wider area,

Fig. 4.3 Examples of biased sex ratios in nature. **(a)** Dock, *Rumex acetosa*: P.D. Putwain and J.L. Harper (1972, *Journal of Ecology*, **60**, 113–29) found a female bias in 15 out of 16 populations sampled. **(b)** Weasel, *Mustela nivalis*: C.M. King (1975, *Mammal Review*, **5**, 1–8) describes how trap density affects apparent sampled sex ratios. **(c)** Ground squirrel, *Spermophilus richardsonii*: S.M. Schmutz, D.A. Boag and J.K. Schmutz (1979, *Canadian Journal of Zoology*, **57**, 1849–55) describe the sources of differential mortality which produce a male biased death rate and female biased sex ratios.

Table 4.4 Sex differences and predation risks in planktonic Copepoda. Male copepods are smaller and more mobile than females. In freshwater lakes they are likely to be attractive prey for visually cued hunters, while females appeal to more sedentary predators seeking larger rewards. A laboratory study by E.J. Maly (1970, *Limnology and Oceanography*, **15**, 566–73) shows that this difference can be significant.

	Sex of copepod		
	Male		Female
Diaptomus shoshone			
Carapace length (mm)	1.73		1.86
Activity index	1.24		1.0
Feeding rate of tadpoles *Ambystoma*	0.32		0.44
Significance of difference		$P < 0.01$	
Diaptomus oregonensis			
Total length (mm)	1.22		1.28
Total width	2.06		2.60
Activity index	1.30		1.0
Feeding rate of guppies *Lebistes*	0.58		0.34
Significance of difference		$P < 0.01$	

particularly during the early dispersive stage as is typical of mammals. The sex ratio of adult corpses collected from debris around hawk nests is similar to that in the local population but among juveniles mortality falls more heavily on males particularly among roadside kills.

In *Diaptomus*, a planktonic copepod, predation of males and females is dependent on the hunting response of specific predators. *Ambystoma* tadpoles assess food items solely by size and are deterred from chasing small or active prey. Consequently they prey more heavily on the larger and less active female *Diaptomus*. Male *Diaptomus* are taken more frequently by the guppy, *Lebistes*, which feeds by grabbing or chasing anything apparently moving: males are more active (Table 4.4). It is possible that the relative abundance of these predators in small ponds could affect the copepod sex ratio.

5 Sex Differences

Male gametes (sperm or pollen) are usually small, cheap to make and produced in enormous numbers. Female gametes (eggs) are relatively large and contain substantial amounts of nutrients. The egg of the ostrich produces over twenty decent-sized omelettes. They can, consequently, be expensive to produce. In addition, in many species of plants and animals the developing embryo is retained by the mother, who protects it and may continue to supply it with food.

These different functions of males and females have led inevitably to the evolution of a diverse spectrum of biological differences between the sexes. Some of these differences are directly associated with their reproductive role. An example is the primary sexual development in mammals. Females have internal ovaries and other organs associated with the growth and protection of the foetus and mammary glands for suckling the young. Males have external genitalia for insemination.

Not all differences between males and females are absolute characteristics of one or other sex. As a consequence of their function many other relative differences between the sexes have evolved where the best strategy for a successful male differs from that for being a successful female. An example of this kind of difference is one of the most commonly observed sexual dimorphisms: size. The evolution of size differences is discussed in Section 5.1. A second kind of sex difference stems from the relative costs of gametes; differences due to sexual selection are discussed in Section 5.2. Thirdly, there are many flexible ecological and behavioural differences between the sexes, particularly, in distribution and feeding; some of the common ecological differences are examined in Section 5.3.

5.1 Size differences

Among the most familiar of animals, the mammals and birds, males are usually larger than females. This pattern among the higher vertebrates is not, however, reflected in the rest of the Animal and Plant Kingdoms. A survey of any large group of animals reveals that in the vast majority of species females are larger than males. A similar pattern occurs among dioecious plants. Animal examples in which females are normally the larger sex include the insects, which are the group with the largest number of species, and the nematodes, which are the most abundant in terms of biomass. Sexual size dimorphism can often be extreme. The female marine worm, *Bonellia viridis* (Section 2.3), is several hundred times the size of the male whilst in some species of deep-sea angler fish the male is but a tiny fused appendage on the body of the female (Fig. 5.1).

In Chapter 3 a range of species were discussed in which individuals had the potential to change sex during their lifetimes. The majority of sex changers among plants and invertebrates tend to be male when small or in poor condition. They become female only when they are large and have good physiological reserves or a plentiful food supply. Big mothers make better mothers and, when the option is available, small individuals are likely to do relatively better as males. Because eggs are more costly to produce than sperm or pollen, female fitness in relation to an increase in egg production will rise more steeply with size than male fitness in relation to an increase in sperm production.

There are many species in which the number of eggs produced by a female does indeed increase substantially with size. In some species there is also selection in favour of relatively small size for males. One factor which may be important in aquatic environments is that smaller males will be more mobile and thus more likely to encounter sedentary females. This explanation is similar to that for the evolution of anisogamy itself (Chapter 1).

Terrestrial insects, which undergo metamorphosis into an adult form of fixed size, also have relatively small males. As the 19th century biologist Alfred Russell Wallace originally pointed out to Charles Darwin, there may be an advantage for a male in emerging as an adult before any females appear. This will permit him eventually to encounter as many unmated females as possible. The phenomenon of early male emergence (which is sometimes confusingly called protandry: see Section 3.2) is well known in Lepidoptera. In some species males will actually seek and guard the pupae of unemerged females or even mate with the pupa itself. Since butterfly eggs are laid without regard to sex in sequence it is argued that males can only be certain of early emergence by changing into adults at an earlier stage in their development, thereby committing themselves to a smaller size. Wallace cited the example of the silkmoths, *Bombyx*, where it is possible to sort male and female cocoons by weighing them: males are lighter. The weight of the cocoon is a rough guide to the time taken to pupate and the lighter males pupate earlier.

Although this is one way in which small male size might be achieved, early emergence is not necessarily a restricting factor on size. In some species of Hymenoptera, such as the digger wasp *Bembecinus quinquespinosus*, males not only guard late-stage female pupae but may dig females out and fly off with them to isolate them from other males. In order to fly with the female the male has to be bigger than her, yet males still emerge much earlier and many males may be present in an area where

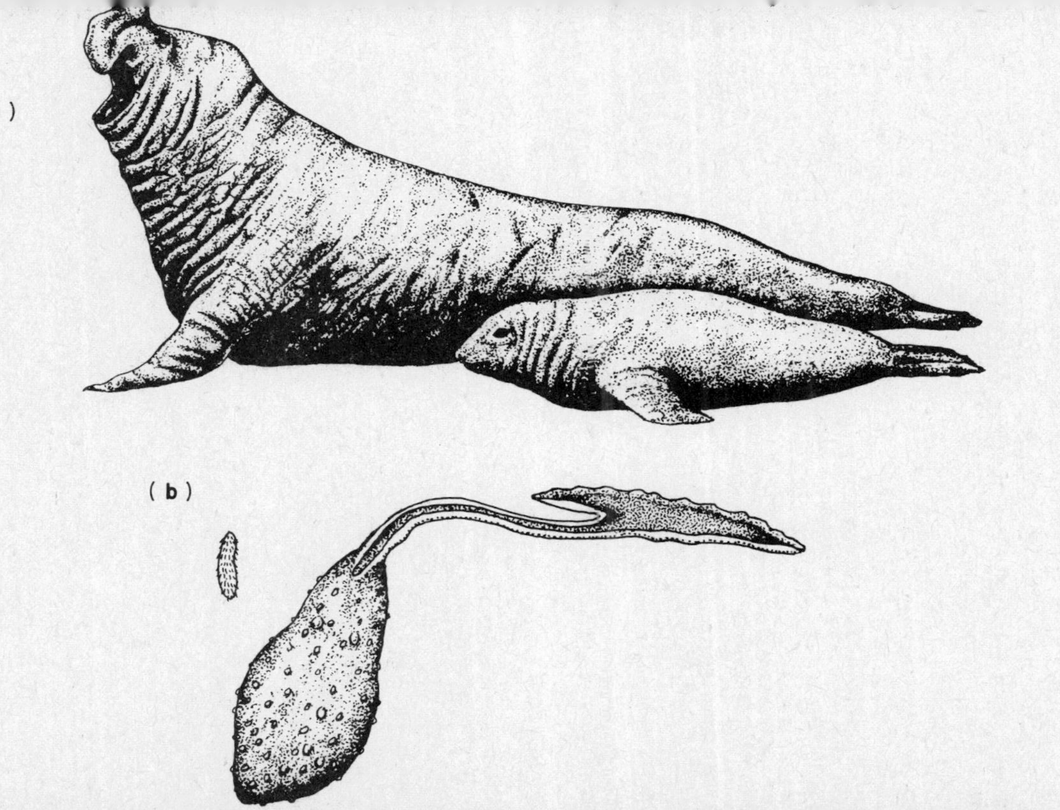

Fig. 5.1 Extremes of sexual size dimorphism in the Animal Kingdom. **(a)** Larger males are common among the higher vertebrates. In the elephant seal, *Mirounga angustirostris*, males may be up to eight times the weight of females. **(b)** Larger females are typical among the invertebrates. Female *Bonellia viridis* may be up to 500 times larger than the male. Several males may live in the gutter of the female's proboscis.

females are emerging. There are also species of aquatic Crustacea, such as sea-slaters, *Ligia*, and freshwater shrimps, *Gammarus*, where the males walk or swim with their mates. In these cases the males are again larger than the females, against the general trend for their taxonomic group.

In the lower vertebrates (fish, amphibia and reptiles), as among invertebrates, males are often the smaller sex. But, as noted above, the trend is for smaller females among higher vertebrates. Males are the larger sex in 20 out of 27 Orders of birds and 16 out of 20 Orders of mammals. Two evolutionary problems emerge from these observations. First, why is the usual trend reversed in these taxa? Second, why are some birds and mammals consistently different? There is at present no entirely satisfactory answer to these general questions. However, there have been a number of hypotheses proposed to account for the 'reversed' dimorphism of groups such as birds of prey and bats.

One possibility is that larger female size has evolved in raptors as a result of a loading constraint. Among the birds of prey there are a range of ecological types and the species can be divided in order of raptorial specialisa-

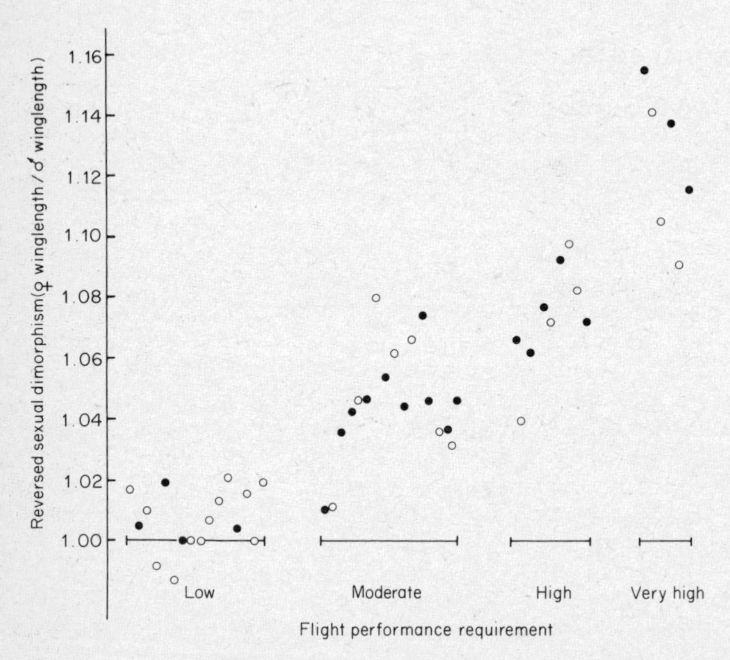

Fig. 5.2 The degree of reversed sexual size dimorphism in birds of prey is related to the demands for flight performance. In species such as the sparrowhawk, which is often an aerial pursuer of other birds, the female is much larger than the male. Each point in the figure represents the average value for a genus: open circle = 1–4 species in genus; solid disc = 5+ species. (Wheeler, P. and Greenwood, P.J. (1983). *Oikos*, **40**, 145–9.)

tion and efficiency from the ones that take rather static prey through to those taking live prey in the air. This last group requires the greatest degree of aerial manoeuvrability with rapid changes in speed and direction. This ability is dependent on wing loading, which is determined by wing area and body mass. During the breeding season the weight of a female sparrow-hawk, *Accipiter nisus*, may increase by up to a quarter. This extra weight is for egg production and fat reserves for the incubation period. One way of decreasing the relative effect of this loading on manoeuvrability would be to increase absolute body size so that the weight increase represents a smaller proportion of her total body weight. If this is a feasible evolutionary solution to the loading problem then it would be predicted that females will be much bigger relative to males in those species with the greatest requirements for aerial ability. The comparative evidence supports this hypothesis (Fig. 5.2). Hawks and falcons which hunt by pursuing birds through the air are the most sexually dimorphic while scavenging ground feeding vultures are the least. Females are larger than males in those skuas, frigatebirds, owls and bats which also pursue and capture prey in flight.

5.2 Sexual selection

One of the most extreme cases of size dimorphism among mammals is the elephant seal, *Mirounga angustirostris* (see Fig. 5.1). Adult males may be up to eight times the weight of females and larger, older males dominate an area of beach where they can monopolise a group of breeding females. Other conspicuous features, such as bright and extravagant plumage among birds or antlers among deer and prominent canines among primates, often occur only in adult males.

When Charles Darwin wrote his book *The Origin of Species* he suggested that the adaptive features of organisms could be explained by their evolution through a process of natural selection. However, he was unable to use the same hypothesis to explain those features which were generally, but not exclusively, possessed by adult males. An example is the extravagant tail of the peacock, *Pavo cristatus*. This bizarre feature is entirely absent from the peahen and is acquired by males only in their maturity. Such a character is difficult to explain in terms of natural selection. Peahens and immature peacocks survive and function with shorter tails and dull plumage. Clearly, an alternative hypothesis is required to explain the peacock's tail.

Since the tail is evidently not an absolute requirement for reproduction Darwin suggested that, instead, it affected the relative quality of the peacock's function as a male. He described this hypothesis as sexual selection in his book *The Descent of Man, and Selection in Relation to Sex*, and defined sexual selection as the advantage certain individuals have over other individuals of the same sex and species, in exclusive relation to reproduction.

It is not surprising that relevant characters are usually seen only in adults since in non-reproductive individuals there is no competition for reproductive opportunities. But why is it that unusual or flamboyant character states are seen in males more often than females? Referring back to the differences in gamete costs, this becomes fairly obvious. As females invest relatively more in eggs than males in sperm, the female's reproductive output will usually be limited by the number of eggs she can produce or care for. By contrast, male reproductive output will be limited by the number of eggs he is able to fertilise. As a result a male is likely to compete with other males for access to females. Females, on the other hand, are likely to be selective about the males with which they mate. The product of both these processes is the differentiation of characters in males which give them an advantage over other males either in direct, male-male competition (Section 5.2.1) or in soliciting female choice (Section 5.2.2).

The differences between natural and sexual selection can be illustrated with the species *Agrion splendens*, a common damselfly. In the summer, males hold territories near sites where females oviposit. When females visit these sites they are courted and guarded by the males. Externally, the sexes differ in the shape of the abdomen and the pattern of the forewings. The shape of the female abdomen is determined by her ovipositor which allows her to place eggs beneath the surface of the water. The shape of the male's abdomen is associated with holding the female during mate-guarding but also with inserting sperm in her reproductive tract. The female's wings are a plain, translucent green whereas the male's forewing is marked by glossy black, blue and white bands. The males only react to females with a particular range of wing types while the quality of his bands may determine his effectiveness during courtship (Fig. 5.3).

In this case, differences in abdomen shape are due to natural selection whereas the wing pattern appears to be derived through sexual selection. What is not always clear, however, is the extent to which a particular character that has evolved through sexual selection is due to male-male (intrasexual) competition or female choice (epigamic selection).

5.2.1 Intrasexual selection

When males are competing it is often the case that the larger individual wins because of its greater size and strength. Such an individual may, as a result, have a higher reproductive success through increased access to females. If male size is heritable then there will be selection for an increase in size until, at some stage, it is checked by the opposing forces of natural selection. This does not necessarily imply, however, that male-male competition will result in the evolution of males larger than females.

Male fruitflies *Drosophila melanogaster* are consistently smaller than females, although there is considerable variance in the size of both sexes. Large males have a greater reproductive success than small males when offered females under various experimental conditions. First, longevity is size-related, so large males have more mating opportunities. Second,

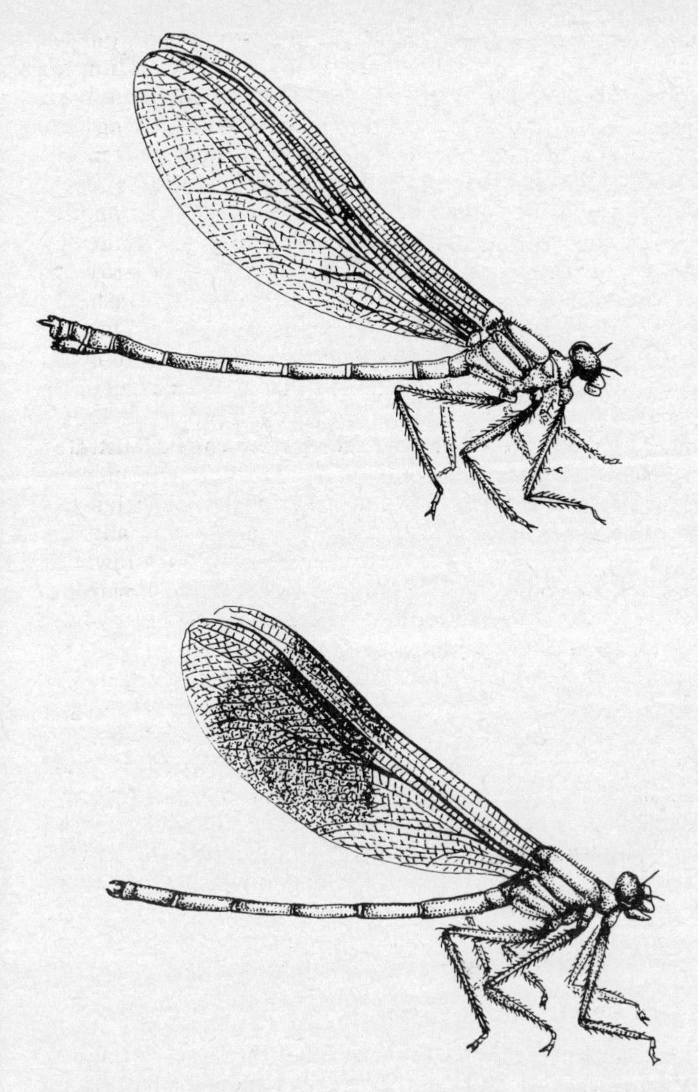

Fig. 5.3 Sexual dimorphism in the damselfly, *Agrion splendens*. Differences in the wing coloration of the female (above) and the male (below) may be due to sexual selection. Differences in the shape of the abdomen (the hind end of the body) are a product of natural selection and are associated with the effective reproductive function of each individual.

larger males mate more quickly. Third, larger males sing (by beating their wings) more often and louder than small males and there is evidence that this is more attractive to females. And, finally, in physical encounters large males always chase off smaller males. Intensive male-male competition also occurs during the short breeding season of the common toad

Bufo bufo and large males can displace smaller ones from the backs of females. Even so, on average females are considerably larger than males.

Amongst the mammals, however, there are examples such as the red deer, *Cervus elaphus*, where such competition between males is correlated with the evolution of elaborate weaponery, in the form of antlers, and larger male size. Studies on the Isle of Rhum have helped to elucidate the factors affecting body size and reproductive success in stags and hinds. Stags compete during the autumn rut to attract and monopolise a harem of up to 20 hinds. The sex ratio is about 0.5 so, consequently, many stags fail to hold a harem. The number of calves fathered is limited by the number of hinds that conceive, but stags which hold larger harems for longer periods have a higher reproductive success. The most successful males father as many as 25 calves during their life, while others never breed. For a stag, the number of calves he fathers is associated with his body condition and fighting qualities. Body condition is related to feeding and social dominance in the previous winter which, in young stags, depends on body weight. Good fighters also tend to be larger animals. The weight of the antlers, again related to body size and weight, is crucial in fights and is closely correlated with lifetime reproductive success. By contrast, although the quality of the calves is related to the hind's body condition, there is no relationship between the mother's skeletal size and the calf's weight.

In polygynous species like red deer, where one male may monopolise many females (see Chapter 6), there is often a range of linked differences between the sexes. First, the variance in reproductive success of males is greater than that of females. Second, species of ungulates, primates and pinnipeds which are the most polygynous are also the most sexually dimorphic, not only in size but also in the development of male weaponry such as antlers in red deer and canines in baboons. Third, sexual maturity is often delayed in species where the male has to be large in order to be successful. And, finally, intensive male-male competition imposes a cost and can result in higher male mortality.

Male-male competition also influences the flowering patterns of plants. In monomorphic flowers, such as *Impatiens* (Section 3.2.1), the protandric phase may be as much as five times longer than the female phase. Milkweed, *Asclepias exaltata*, is another hermaphroditic plant and it produces far fewer seeds than it originally has flowers. Female reproduction seems to be limited by resources because most young fruits are aborted. In an experimental study in Michigan, plants were modified either by having half their flowers removed or by having pollinators excluded for half the flowering time. Nonetheless, experimental plants produced as many fruits as did unmodified controls. These results suggest that the number of flowers and the time for which they are open must be traits selected to enhance the effectiveness of male function in pollinating other plants rather than individual seed production.

Intrasexual competition is much less common among females: male gametes are rarely a limiting resource. However, gametes are not the only cost when breeding and there are circumstances in which male investment may exceed the female's. There could then be competition among females for access to males, which would be exacerbated if males varied in quality and a high quality male enhanced reproductive success. Such a condition appears to occur among moorhens, *Gallinula chloropus*.

Female moorhens are more aggressive than males and initiate courtship more frequently when pair formation occurs in winter feeding flocks. A paired female will physically attack another female if the second bird approaches her mate. In such encounters the females risk injury by clawing at one another but heavier birds are more likely to win. What advantage does a female gain from such antagonism? The male moorhen performs most of the incubation and larger females tend to be paired with males that have the largest fat reserves. This allows him to spend more time on the nest and such pairs make more breeding attempts in a season.

5.2.2 Mate choice

Because eggs usually are a limiting resource, while male gametes are abundant, a female can rarely increase her reproductive potential by mating with a number of partners. She may, however, be able to increase her ultimate success and have more surviving offspring if she is selective about the male gametes which fertilise her eggs. Such selection could occur at a number of stages before and after sexual contact. Among animals, behavioural cues may allow a female to discriminate among males who court her. In both plants and animals, male gametes can be selectively prevented from fertilising eggs after pollination or insemination. And, even after fertilisation, selective abortion can give the female an opportunity to choose which embryos receive further investment. This final option is likely to be less common, however, since it is certainly more costly and may reduce reproductive potential.

Many pollen fall on the stigma of a flower but not all germinate and some tubes only partially develop. Biochemical tests in the female tissue (the style) can act as an effective sifting mechanism to weed out particular genotypes among the available males. In maize, *Zea mays*, selection has a clear effect on fitness: pollen tubes which bear genotypes producing larger seeds and seedlings appear to be favoured.

Mate choice among animals is often more overt, yet it is not always easy to understand the reasons why some male characters should necessarily be correlated with the bearer's quality and why this should be a reliable criterion for female choice. For example, the sedge warbler, *Acrocephalus schoenobaenus*, has an extremely complex song. Males with the most complex repertoire pair sooner in the breeding season than those with simpler songs and as soon as they have obtained a mate they stop singing. It is assumed that females are choosing males on the basis of their vocal quality: it is an acoustic equivalent of the peacock's tail. How could a female which selects a mate with a particularly large song repertoire or bright plumage actually be improving her fitness? Initially, perhaps, females may have used superficial features in making a choice of males as an indication of their fitness. For example, males with slightly more intensive displays, brighter colours or elaborate songs may have been accurately indicating a superior quality over other males. If females start selecting on the basis of these traits then eventually such characters can evolve in isolation, decoupled from the original fitness quality. Mothers choosing brightly coloured males will tend to produce daughters which also prefer bright males and sons which are bright. This preference may result in the rapid evolutionary exaggeration of the preferred character.

The long-tailed widowbird, *Euplectes progne*, of central Africa has males which possess long and extravagant tail feathers during the breeding season. Males compete for territories in areas of open grassland into which they attract a number of females by a slow aerial display involving the expanded tail. Several females may nest within a territory and raise their young unaided by the male. The question is whether the long tail of the male, which is absent from the female and from immature males, is used as a signal in male-male competition for territories (females later choosing males with high quality territories) or whether the tail is the product of female choice in courtship display.

Field experiments involved a study of male success in attracting females before and after their tail length had been manipulated. Birds were studied to determine their baseline mating success and then snared and treated in four experimental groups. Control birds were released without modification. A second control group had their tails cut but reset at the same length with a fast-drying glue. No change in mating success would be predicted in these groups, apart from that due to the trauma of being handled. A third, experimental group had their tails cut and restuck at a new, shorter length than before, while a final group received the section cut from the third group so that their tails were extended by about half. Here, it is predicted that docking will reduce and extension will increase mating success. It is worth noting that when birds were released back into the bush their territory size was found to remain unchanged. This suggests that tail length played no role in interactions between males, but it would be necessary to check on the effect of tail docking during territory establishment to confirm this. It is clear, however, that birds with super-long tails had an enhanced mating success. Thus, females appear to favour males with long tails (Table 5.1).

Why don't male widowbirds have super-long tails naturally, if they are such a help in getting a mate? The problem of why this sort of male character does not run away to even greater excess has been posed by several evolutionary biologists. The simple answer is probably that natural selection intervenes to impose some limit at which the benefits gained through sexual selection are checked by the penalties of having to carry the appendage around during normal activity. A long tail may hamper flying or make it difficult for the bird to balance while feeding, whilst its production and maintenance will entail considerable metabolic costs.

Bright coloration may signal quality in some species in a more direct fashion. Because of the cost of producing bright plumage, only those indi-

Table 5.1 Female choice in widowbirds, *Euplectes progne*. Male widowbirds have very long tail plumes that are absent in females and immature birds. If long tails have evolved through female choice then we should predict that females would selectively pair with the longest tailed males. Experimental manipulation shows this to be the case (M. Andersson (1982). *Nature*, **299**, 818–20).

	Mean nests per male	
Experimental group	Before	After
Males with tails shortened to 14 cm	1.33	0.44
Males with tails cut but left at 50 cm	1.55	0.89
Males with uncut tails around 50 cm	1.44	0.44
Males with tails elongated to 75 cm	1.67	1.89

viduals which are in good condition can afford the energy and materials involved. W. D. Hamilton and M. Zuk suggested that males which are in poor condition, due for example to a heavy parasite burden, would produce a poor display which females would reject when faced with showier, brighter suitors. Similarly, the high quality males would be able to court more vigorously and for longer. Thus external signals could be an immediate index of the condition of a male and his resistance to disease: a quality worth passing on to offspring.

The relative costs of gamete production mean that females are usually the choosy sex but there are circumstances in which incidental costs may make it desirable for males to be selective about the females with which they mate. If a male carries a female during precopula guarding, as some Crustacea do, then the energetic costs involved, particularly when swimming, may mean that in the long term it would be advantageous to pair with a smaller female despite their lower fecundity. The special case where males rather than females invest time in parental care and should therefore be selective – this has been described in birds such as phalaropes and jacanas – is also discussed in Chapter 6.

Another case of male choice occurs in the marine crustacean, *Pseudosquilla ciliata*. Females brood the eggs but also initiate courtship significantly more often than males and persist in inducing coy males to copulate. In a laboratory arena females were seen to be opportunistic in their approach and actually chased and attempted to grasp fleeing males with their maxillipeds. Such an aggressive approach was rarely successful. Males by contrast are a cautious and discriminating sex and are significantly more likely to choose to copulate with females larger than themselves rather than with partners equal in size or smaller. As in other Crustacea, large female *Pseudosquilla* carry more eggs.

There are two reasons why males might be choosy in circumstances where apparent costs seem to be small. First, because female *Pseudosquilla* are territorial and more aggressive than males they tend to escalate aggressive encounters. This can and does result in actual injury to males, hence their negative response to over-ardent courtship. Second, females mate frequently and with a variety of partners, producing a significant level of sperm competition in their spermathecae. Males should then be conservative about using sperm which, though cheap, may take some time to replace. Males would increase their reproductive potential by saving sperm for larger clutches, as observed in experiments.

Male red-spotted newts, *Notophthalmus viridescens*, also discriminate in favour of larger partners. Males spend significant periods courting females and the amount of sperm available is limited, but the number of eggs laid after insemination is closely related to female body size. Surprisingly the male newt does not have to see his potential partners to make his choice. In a Y-maze experiment, where males were simultaneously exposed to water draining from two tanks each with a single female, they consistently showed a preference for the maze arm with the larger animal. Whether this reaction is caused by the chemical nature of the female's odour or by its quantity is as yet unknown.

There are a number of species where the sexes are similar in both size and external appearance. These include the large, long-lived birds with lengthy incubation and fledging periods where it is usually essential that both parents are present to raise any young successfully. Examples include the gannets, *Sula*, the divers, *Gavia*, and the grebes, *Podiceps*. Here the sexes have a similar plumage, often brighter in the breeding season than in winter, and indulge in mutual courtship. The penguin dance of the great-crested grebe, *P. cristatus*, is as much dependent on the female displaying to the male as the other way round. In the past such behaviour has been interpreted in rather vague terms, such as 'cementing the pair-bond' or 'synchronising the breeding cycle'. A more evolutionary approach would suggest that, where dependence on the partner is critical, mutual courtship enables both male and female to assess the quality of their prospective mate. Often associated with this is the appearance of sexually selected characters in both sexes during the spring and the big, bright bill of the puffin, *Fratercula arctica*, is a delightful example.

5.3 Sex differences in ecology

One of the consequences of sexual dimorphism is that the ecology of males and females may also differ. Some of the differences in behaviour may be very flexible. The chaffinch, *Fringilla coelebs*, is one of the commonest breeding birds of northern Europe. Males are brightly coloured and slightly larger than the brown females. In the breeding season birds pair up on territories defended by the male and both sexes contribute to offspring care. In winter, however, chaffinches form feeding flocks and, in continental Europe, such flocks are usually sexually segregated. Females also disperse further south than do males. There are a number of reasons for this separation. Males tend to remain relatively close to the area where they seek to hold territories. The smaller females have no need to remain near the breeding ground, but if they did they would suffer from competition with males and also lose body heat faster. Thus they can take the opportunity to make an advantageous move to a milder climate. A similar seasonal segregation by sex is seen in the red-billed quelea, *Quelea quelea*, in Ethiopia.

The distribution of males and females also differs among mammals. In Section 4.4 it was noted how trap density affects estimates of sex ratios among weasels. This is caused by their different ranging patterns. A female has a small home range centred around a retreat where she can hide and care for her young. A male has no such ties and ranges more widely across an area that includes a number of potential mates. Such a pattern also occurs in the red fox, *Vulpes vulpes*. Female foxes have ranges up to 5 km^2 whereas male ranges can cover as much as 15 km^2

Habitat use may differ even within a territory occupied jointly by a pair of animals. Male and female yellow-rumped warblers, *Dendroica coronata*, use the same kinds of feeding perches but feed in different trees and at different heights. Males forage in taller trees at a greater distance from the ground; this allows them to stay close to their song posts. Females spend a greater proportion of their time near the nest-site and are more often seen close to the ground. Their distributions tie in closely with their different reproductive roles.

Spatial separation of the sexes occurs among a number of species of dioecious plants. The sex ratios of populations of both salt-grass, *Distichlis spicata*, and of meadow rue, *Thalictrum fendleri*, vary significantly along transects taken across habitat gradients (see Section 4.4). Males are clustered near the saline end of the salt-grass series and in dry, sunny sites on the meadow transect. Other species have similar non-random distributions, particularly with respect to soil moisture. There are two reasons why this might be so. First, males are smaller and less sensitive to water stress than females and may survive better in dry sites. Second, pollen dispersal is much higher from dry and sparsely vegetated habitats than among dense stands on moist soil. By contrast, seed production is enhanced by good growing conditions over a longer period. The plants may control sex expression to be male in dry habitats and female elsewhere.

As well as differences in distribution there are feeding differences between male and female animals in sympatric conditions. Among invertebrates the smaller males may cease feeding entirely once mature and devote their time instead to locating receptive females. Females, by contrast, benefit from continuing to feed and to produce more eggs. The nutritional requirements of males and females differ in quality as well as quantity. Female *Calidris* sandpipers in tundra habitats are reported to feed opportunistically on fragmented small mammal bones. This provides an essential supplement to their calcium-poor diet which is invested in egg yolk and shell.

The large red deer stags satisfy their higher nutritional demands by grazing in areas with abundant but poor quality vegetation. Quality is

Table 5.2 Sexual size dimorphism on islands. Ladder-backed woodpeckers are found in communities including several woodpecker species on the North American mainland but usually only one species is found on each of the Caribbean islands. Where there are fewer competitors the feeding niches of males and females have diverged (Selander, R.K. (1966). *The Condor*, **68**, 113–51).

	Male	Female	% difference
Dominica: *Melanerpes striatus*			
Weight (g)	83–92	69–75	18.1
Bill length (mm)	30–37	22–29	21.3
Length of horny tip of tongue (mm)	15.4–15.7	10.2–10.4	34.5
Texas: *Melanerpes aurifrons*			
Weight (g)	73–99	66–90	10.5
Bill length (mm)	30–37	28–34	9.1
Length of horny tip of tongue (mm)	14.6–16.1	12.2–15.0	9.4

sacrificed for quantity, but they may compensate for this by being able to process a greater volume through their digestive system. Size dimorphism often leads to secondary feeding differences. Darwin noted that male goldfinches, *Carduelis elegans*, can feed on teazle, *Dipsacus*, seeds while females feed more commonly on betony, *Scrophularia*. The marginally longer beak of the male allows it access to a food source the female is unable to reach.

In territorial species, a male and female of a pair could potentially compete for food with one another more than with other members of their species. By exploiting different niches within a territory they can minimise the amount of competition between them and increase the range of resources both for themselves and any dependent offspring. One of the most remarkable examples was the huia, *Heteralocha acutirostris*, of New Zealand, now extinct. The male had a straight, stout bill for, presumably, extracting grubs buried in rotten wood while the female had a long, curved bill for poking into existing holes and crevices in trees for insects. In fact ecological divergence of the sexes is often a feature of impoverished faunas typical of isolated or island communities. In the absence of intense interspecific competition species can exploit a wider niche and instead of both sexes overlapping in diet the expansion in food availability results in a divergence of their foraging behaviour and prey selection.

This seems to have occurred with a number of North American species of ladder-backed woodpeckers. On the mainland, species such as the golden-fronted woodpecker, *Melanerpes aurifrons*, which is found from Texas southwards, shares its range with a number of other species. Its plumage is sexually dimorphic but while males average a little larger than females there is extensive overlap on body and bill dimensions. At a similar latitude among the islands of the Caribbean a group of closely related *Melanerpes* woodpeckers are found, but each is endemic to a particular island. The Hispaniolan woodpecker, *M.striatus*, for example, is strongly size dimorphic. The male has similar bill dimensions to *M.aurifrons* but the female's bill is only about three-quarters as long and there is little overlap between measured specimens of either sex (Table 5.2). The male's bill is also stouter and whereas male *M.striatus* spend significantly more time probing and excavating, the females spend more time in gleaning insects from wood surfaces.

6 Mating Systems

Traditionally, the nature of the link between males and females has been the basis for the classification of mating systems. ***Monogamy*** is the term for a one-to-one association between the sexes during the reproductive period. This may, in some circumstances, extend throughout the whole breeding period as a pair bond. In ***polygamy*** a single member of one sex has access to more than one member of the other sex. The nature of this access can vary from the fleeting to the long term. Familiarity between the sexes is counted in seconds for midges, in years for gorillas. When, in fact, access is non-exclusive, as it is in many invertebrates, we might regard the system as ***promiscuous***: an individual of either sex may mate with a series of partners of the other sex. Polygamous mating systems often occur where a single male is able to obtain exclusive reproductive access to a number of females, either simultaneously or serially. This is correctly described as polygyny. Where, by contrast, a single female has exclusive access to more than one male the system is described as polyandry.

Recently, alternative ways of classifying mating systems have been devised. Some entail a more precise description that includes additional information about the social organisation. A spectrum of social types can be found within groups such as, for instance, the primates. The breeding unit of the arboreal gibbons *Hylobates* of south-east Asia is a monogamous one consisting of a father, mother and one or two dependent offspring living within an exclusive territory. Alternatively there are polygynous single male groups, such as those of the large terrestrial Hamadryas baboon, *Papio hamadryas*, which lives in large herds roaming through rocky desert. Each troop is divided into bands and sub-units where one male controls a harem of up to 10 adult females. In west African forests, by contrast, multi-male troops of the red colobus, *Colobus badius*, contain a mixture of 40–60 young and adults of both sexes. Some caution is needed in classifying species to particular categories in the absence of detailed field observations. Gorillas live in multi-male/multi-female cohesive social units, but in terms of reproduction they operate as single male groups since only the dominant, silverback male mates with the females.

Another type of classification gives an indication as to the means by which males and females actually acquire mates. The two commonest mating systems among higher vertebrates involve either resource defence or mate defence. In the former type, one sex (in most species the male) defends a resource such as a feeding territory or nesting site in order to attract a mate. In the latter a male is more concerned with defending and guarding a female than acquiring the resources which may actually determine their spatial distribution. Both types have the capacity to result in either monogamous or polygamous breeding systems.

A species' mating system will have a number of implications for the amount of time and effort that the parents invest in their offspring. Unfortunately, one of the major problems at the present time is the lack of understanding of the factors which determine what type of mating system occurs and how much each parent should invest in their young. Most species of birds are monogamous and the male often assists the female in the rearing of the young. By contrast most mammals are polygamous with low paternal investment. These broad patterns will first be explored through specific examples, followed by a more detailed look at the reasons for the diverse range of avian mating systems.

6.1 The mating systems of birds and mammals

One of the most obvious biological differences between birds and mammals is their mode of reproduction. Birds lay eggs which are incubated at a fixed nest site. Some birds, such as waders and waterfowl, produce young which are well developed (precocial) at hatching and are able to leave the nest to feed soon after. In other species, such as the songbirds, the young are poorly developed (altricial) and remain in the nest until fledging. For precocial birds the primary parental duties on top of incubation are probably to protect the offspring from predation and adverse weather conditions when the chicks are small. The young of altricial species are dependent on their parents for the delivery of food to the nest. In both cases, the need to incubate eggs and the extended protection or feeding of the young would imply a relatively high potential for both maternal and paternal investment.

The dominant features of mammalian reproduction are female gestation and lactation. In other words, for a considerable period during a breeding cycle the protection of the embryo and foetus and the feeding of the young seem of necessity to be the sole responsibility of one sex. The potential for male investment in the offspring is much reduced. This may indeed free the male, to some extent, to search for other potential mates. Even in those species such as the monogamous marmosets (Callitrichidae), where there is paternal assistance in carrying one of the babies, the male continues to return the infant to its mother to be fed.

There are a number of other factors which are linked to this difference in paternal investment between the higher vertebrates. For example, there is a striking contrast between, on the one hand, the evolution and use of weaponry and the perfunctory courtship of male mammals and, on the other, the ritualised displays and bright coloration that advertise male birds. In mammals, the low level of paternal investment is predictably a feature associated with polygynous mating systems. Polygyny is more likely to be achieved by a male's success in defending a group of mates than through resource defence. Highly polygynous species tend to exhibit high levels of male-male aggression and there may be limited scope for female choice (Chapter 5). In birds, high paternal investment is a feature of monogamous mating systems. The system arises through resource rather than mate defence and, whilst intrasexual competition is frequent, there appears to be much more scope for mate choice amongst birds than amongst mammals.

Despite these associations we are a long way from an understanding of the evolution and consequences of mating systems. One approach which may elucidate some of the facts is to outline a series of life history traits for birds and mammals and then see to what extent some of these traits are fundamental causes of particular mating systems and others mere consequences. To emphasise the complexity of the problem, some of the major points of difference between a monogamous bird and a polygamous mammal are summarised.

Female red deer, *Cervus elaphus*, and any dependent offspring live within a home range in large groups. During the rutting season there is intense male-male competition in the form of roaring contests and fights. Successful stags defend a group of hinds against other males. The mating system is one of polygynous mate defence: females appear to have little say in the matter. When the female gives birth in the spring she leaves the fawn in hiding and returns to suckle it infrequently to minimise the risk of predation. Any female young remain on their mother's home range as a member of a matrilineal social group, while young males leave their natal home range to join other wandering stags.

There are a number of ways of looking at this system which can lead to questions about its evolution. For example, female gestation and lactation limit the potential for investment by stags. Does this predispose the males to polygyny and to mate defence? Alternatively, females are dispersed in large groups which require extensive home ranges. Does this limit the capacity of a male to defend an appropriate feeding resource? This might result in polygyny through mate defence. Would the limits to male investment then reinforce the selection for maternal responsibility for raising the young? Clearly different arguments can be used to approach the problem from either side.

In blackbirds, *Turdus merula*, the breeding system is very different. A single cock blackbird establishes a feeding territory at the start of the breeding season. He is open to selection or rejection by females which visit him. Eggs, once laid, are incubated at a fixed nest site, the young are altricial and they are fed frequently throughout the day by both parents. The mating system is one of monogamous, resource defence and, in most birds, it is the female offspring rather than the male which are likely to end up moving and breeding further from home.

The presence of altricial young in a nest increases the potential for high male investment. Does this lead to the acquisition of feeding resources by the male and to monogamy? Alternatively, reproductive success for a blackbird depends on a good nest site and a high quality territory. Does this lead to resource defence by the male, which limits his capacity to attract more than one female? Would this predispose him then to invest in those young he has fathered?

Avian mating systems have received most attention to date. More insight into likely evolutionary pathways can be gained by looking in more detail at some of these.

6.2 Avian mating systems

6.2.1 Monogamy

About 90% of bird species are monogamous. There are a number of different ways in which such a mating system might have evolved. Many large, long-lived birds have long incubation periods and young which are dependent on their parents for some time after hatching. For such species, like albatrosses and gulls, both male and female share incubation duties and after hatching both feed the young. It is probably essential for both parents to be present in order for either to have any reproductive success. If one deserted or died it would be extremely difficult for the other to carry on alone. To leave the nest exposed for long periods increases the risk of the eggs or young fledglings becoming chilled or being taken by predators. On the other hand, to stay at the nest to incubate the eggs or brood the young increases the parent's chances of starvation.

Other birds, particularly the passerines, do not have a system where it is essential for both parents to be present to raise some offspring. Under these circumstances there will be a conflict of interest between the sexes. For a male, it is advantageous to mate with as many females as possible. But a female should preferentially pair monogamously so that the male assists her in rearing the young.

Imagine an hypothetical species where the parents together can successfully raise, on average, six young to fledging but where a female alone can manage to feed only three chicks. It would be to the male's advantage if he could induce two or more females to nest in his territory. By helping one mate and ignoring the second the male could potentially father nine offspring. It would be to the second female's advantage, however, to be the sole breeder in the territory so that she can produce six young with the male's assistance. The reason that most birds are monogamous is probably because the male cannot defend a territory of sufficient size or quality to attract more than one female. But it is worth stressing that little attention has been paid to why this sort of territorial system evolved in the first place.

Finally, there are a number of monogamous species such as ducks where male investment in offspring is nonetheless minimal. For example, male eiders, *Somateria mollissima*, will stay very close to a female at the start of the breeding season. This mate guarding continues until a few days after the female has started to incubate the eggs. The male then deserts and leaves her to complete the incubation and, eventually, to lead the ducklings to the sea. In this case monogamy is effectively imposed on males by the synchronous nesting of the females. By the time the male deserts his mate the vast majority of other females in the population will also be incubating eggs, so denying males any additional mating opportunities.

6.2.2 Polygyny

Nearly 10% of species of birds are polygynous and there are two main types of avian polygyny. One derives from a high variance in the quality or quantity of resources which males defend in order to attract females. The second involves no defence of resources. Instead males congregate at a traditional display site, known as a lek, and females choose a mate from the assembled collection.

Resource defence polygyny occurs predominantly amongst those species which live in patchy environments, such as grasslands or marshes. Some males can monopolise high quality sites while the remainder are restricted to areas of lower quality. One of the most detailed studies of such a system is that of the lark bunting, *Calamospiza melanocorys*, a North American passerine. The males compete for territories on arrival at their breeding grounds in spring. Some males are able to attract two females, even though there are unpaired males with territories. Each successful male only helps the first female which settles and thus the breeding success of the second is reduced. Why, then, does the second female choose to mate bigamously rather than to be monogamous elsewhere? The critical factor is the amount of cover available at a nest site. The overheating of young in unshaded nests is a major cause of mortality. As a result, females on territories with significant amounts of natural or artificially provided cover have a higher reproductive success than females nesting on territories with limited cover (Fig. 6.1). What is not clear, however, is why males with the best territories do not attract more than two females.

One condition that underlies the system found in lark buntings and other polygynous species is that when secondary females make a choice they should be aware of their mating status. In an unusual instance of polygyny this appears not to be the case. The pied flycatcher, *Ficedula hypoleuca*, is a summer migrant to Europe which nests in natural holes or artificial nest boxes. A proportion of the males defend two territories and these are often widely separated. The male will concentrate his efforts on the second territory once a primary female has settled onto the first territory and started to incubate. If he succeeds in attracting a second female she will be duped into mating on the basis that she is forming a monogamous pair bond. But once the eggs have been laid, by which time it

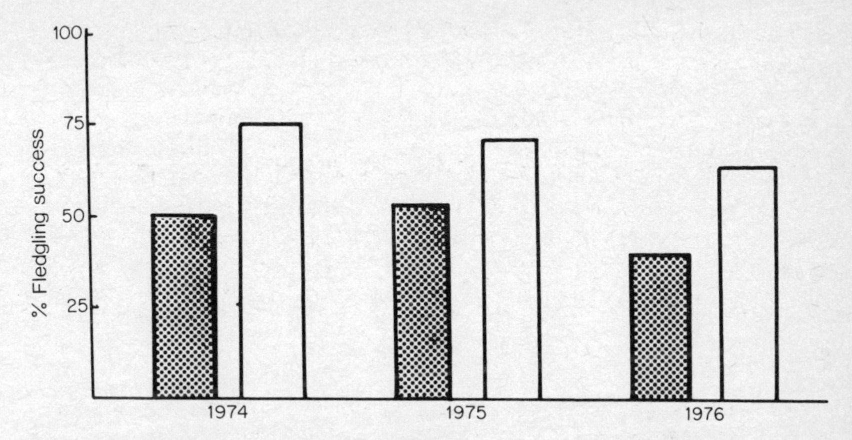

Fig. 6.1 Female lark buntings, *Calamospiza melanocorys*, appear to increase their reproductive success by nesting on territories with plenty of cover to shade nestlings, even where their mate is already paired with another female. The correlation between natural nest cover and fledging rate is highly significant and nest cover can also be manipulated artificially to show that extra cover does indeed increase fledging. Shaded blocks = control nests; open blocks = nests with extra cover. (Pleczynska, W. (1978). *Science*, **201**, 935–7.)

is too late in the season for her to find a new mate and start again, the male returns to feed the brood of the first female. As clutch size declines rapidly during the season, the primary female's reproductive success is likely to be higher with male assistance than the second female under similar circumstances (Table 6.1).

The system of lekking at a communal site has evolved at least twenty times in birds and includes species of grouse, waders, birds of paradise and manakins. It has also been observed in insects, such as chironomid midges and Hawaiian *Drosophila*, and among coral-reef fish and some species of

Table 6.1 Polygyny in pied flycatchers, *Ficedula hypoleuca*. Some male flycatchers are able to pair with more than one female, but desert their second mate once she has laid and return to assist in feeding the primary female's brood. This has an effect on the reproductive success of the two females (Winkel, W. and Winkel, D. (1984). *Journal für Ornithologie*, **125**, 1–14).

Female group	Number studied	Average age (yrs)	Clutch size	Fledged young
Primary female with male assistance	313	2.67	5.64	4.65
Secondary female alone	61	1.67	5.48	2.90

Secondary females are able to raise fewer young when working alone. On average the second broods hatched seven days later than the first.

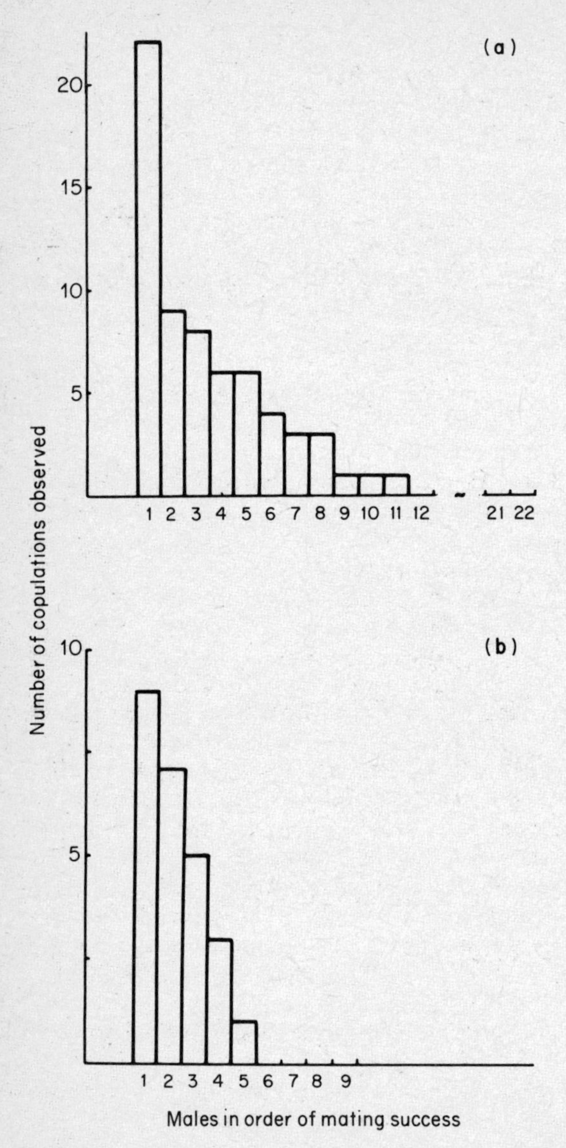

Fig. 6.2 The reproductive success of male birds and mammals on leks (ritualised courtship areas) is highly skewed and some males achieve many matings while others achieve few or none. **(a)** Uganda kob, *Kobus kob thomasi*: 64 matings were observed on a lek occupied by 24 male antelope. (Floody, O.R. and Arnold, A.P. (1975). *Zeitschrift fur Tierpsychologie*, **37**, 192–212.) **(b)** Black grouse, *Lyrurus tetrix*: 25 matings were observed on a lek occupied by 9 males. (Kruijt, J.P. and Hogan, J.A. (1967). *Ardea*, **55**, 203–40.)

mammals, including the Uganda kob *Kobus kob thomasi* and the hammerheaded bat *Hypsignathus monstrosus*. At a lek males are frequently aggressive towards other males and manoeuvre to obtain optimal position. Males are often much larger than females and, amongst birds, they often have bright plumage. Males display intensively at a female when she visits the lek site and she eventually chooses just one of the males with whom to mate. Females tend to choose the same male independently of each other and often this is the male at the centre of the lek. Consequently the distribution of matings is heavily skewed towards certain individuals (Fig. 6.2). The male's sole investment in offspring is sperm, while the female alone is responsible for all aspects of rearing the young.

The ecological factors which favour the evolution of leks are poorly understood. Is a male unable economically to defend a feeding territory, unlike many other birds, or have leks evolved because of a female preference for assessing a range of males? It is not known, in fact, why females select the same males. Do females choose particular males because of their appearance, behaviour or position on the lek? If their choice is based on qualitative differences between males are these the sort of traits which could subsequently be passed on to their sons?

6.2.3 Polyandry

Less than 1% of birds are polyandrous, making this the rarest type of mating system. A female may form a pair bond simultaneously with several males or with a sequence of partners. The predominant pattern found in the Animal Kingdom, of males competing for mates, is thus reversed and females compete for access to males instead. In such species females are usually more aggressive than males and have a brighter plumage.

Some indication of the probable evolutionary transition states from monogamy to polyandry can be traced through the Arctic waders (Charadriiformes). Females are larger than males amongst most of these species. This may be a consequence of the need to start egg laying immediately on arrival at the breeding grounds where the breeding season is very short. As a result, there may be a requirement for putting on extra weight for egg production and incubation during the migration period. Most of these species are monogamous and males share incubation duties with the female. In species such as the sanderling, *Calidris alba*, and Temminck's stint, *Calidris temminckii* (Fig. 6.3b) females lay two clutches. She produces and incubates the second while the first is being incubated by the male. In the dotterel, *Eudromias morinellus*, only one clutch is produced which is incubated by the male alone. And in a few species, such as the red-necked phalarope, *Phalaropus lobatus*, and spotted sandpiper, *Actitis macularia*, the female produces clutches for a series of males.

The pattern may have evolved because of the high energetic cost to the female of rapidly producing a clutch after arriving at a breeding site. She

Fig. 6.3 Mating systems in birds. **(a)** Ruff, *Philomachus pugnax*. Females visit male display grounds (leks) to mate. They alone incubate and care for young. **(b)** Temminck's stint *Calidris temminckii*. Females produce a clutch which is incubated by the male. They then lay a second which they incubate. **(c)** American jacana *Jacana spinosa*. Females mate with several males and lay a clutch for each of them to incubate.

would then have few reserves left for incubation and would need to be relieved by the male for long periods. An alternative condition for the evolution of polyandry would be where the female needed to be able to produce replacement clutches rapidly as a response to high levels of predation. This appears to have been the case in a tropical American waterbird. Female *Jacana spinosa* (Fig. 6.3) defend a large territory against other females but several males may hold smaller territories encompassed by a single female's. The female jacana mates with each male and lays a clutch for the male to incubate. Predation levels are high and the female, relieved of incubation duties and free to feed, is able to replace clutches rapidly when the eggs have been taken by a predator.

6.3 Parental care

A number of species where the type of mating system is intimately linked to the pattern of parental care have already been discussed. For example, those species in which both parents are essential for the rearing of young are predictably monogamous. But, as previously noted, establishing the causal link between mating systems, parental care and other life history features is not always so direct.

A survey of the Animal Kingdom reveals that the amount of time and energy that the two sexes devote to the rearing of offspring varies enormously between species. In most species there is no parental care. In others either the male or the female looks after the young. Finally, in some species, both sexes care for the offspring (Table 6.2). Among the Arthropoda, the absence of parental care is by far the most widespread pattern. Maternal care is well represented in the spiders (Aranea) where the female provides a protective covering for the eggs and may carry the young around, as in the Lycosidae. Among many Crustacea, such as woodlice, eggs are brooded in a marsupium. The provisioning and protection of potential queens by sterile female siblings is a feature of the social Hymenoptera, while paternal care is found in harvestmen (*Opiliones*, called daddy-longlegs in America). Finally, biparental care, the rarest type, is seen in dung beetles, *Scarabeidae*, and the termites (Isoptera).

Table 6.2 An assessment of the relative incidence of the four possible types of parental care in different groups of animals (1 = common, 4 = rare).

	No care	Female	Male	Both
Invertebrates	1	2	3	4
Fish: non-teleosts	2	1	3	–
teleosts	1	3	2	4
Amphibians	2	1	3	4
Reptiles	1	2	4	3
Birds	–	2	3	1
Mammals	–	1	–	2

(a)
(b)
(c)

Within particular taxonomic groups there are a number of life history differences between those species with no parental care and those with parental care. The species which do have parental care tend to have a slow rate of development and relatively large adult body size; a period of immaturity followed by repeated reproduction; long life span and slow population increase; and, often, a well developed social organisation.

Habitat differences can also affect the incidence of care. Freshwater fish are more likely to guard their eggs than marine species. The marine environment is relatively homogeneous in terms of physical and chemical conditions, so there may be little advantage in caring for eggs through maintaining them in more or less favourable locations. The mobility of adult marine fish, such as the cod, *Gadus gadus* (Fig. 6.4a) militates against extended care while appropriate feeding sites of adults and planktonic immatures may also differ. Conversely, there is extreme local variability in physical and chemical conditions in freshwater rivers and lakes. Parental care may have evolved to protect eggs from predators and to restrict their location to the most favourable areas. A structured habitat, among rocks on a river floor, may also make this more feasible.

Among those fish species which do exhibit parental care it is usually only one sex, more often the male, that guards the eggs and young. Fish, unlike most birds and mammals, do not provide food for their young. Care usually consists of fanning the eggs to maintain a high supply of oxygen, cleaning them of parasites, removing diseased eggs and protecting the eggs against predators. Such duties can readily be performed by a single parent. When the species has internal fertilisation eggs are obviously fertilised by the male before they are laid. The female is then likely to be left to guard the eggs because the male is given an opportunity to desert before they are released. When fertilisation is external the sperm do not fertilise the eggs until after they have been laid by the female. In this situation it is more likely that the male will be left with parental responsibility whilst the female has the chance to desert. For example, fertilisation is external among the sea-horses, *Hippocampus* (Fig. 6.4b) and males have a special pouch on their tail to brood and provision the eggs and young.

Although there does appear to be a link between the mode of fertilisation and the pattern of parental care, the association may not be as straightforward as it first appears. Species with external fertilisation also tend to have male territoriality. If there is variation in the quality of territories then females should prefer to lay their eggs on high quality ones. Male care may therefore have evolved as a direct result of resource defence and not necessarily through the mode of fertilisation. One of the most

Fig. 6.4(Left) Parental care in fish. (**a**) Cod, *Gadus gadus*. Typical of most marine fish, fertilised eggs are free-floating members of the plankton and there is no parental care. (**b**) Sea-horse, *Hippocampus* sp. The male has a vascularised pouch in which he carries and nourishes the fertilised eggs. (**c**) Three-spined stickleback, *Gasterosteus aculeatus*. A male builds and protects a nest where he cares for the eggs produced by one or more females.

familiar species with this pattern is the stickleback, *Gasterosteus aculeatus* (Fig. 6.4c). The male builds a nest of plant material bound with mucus. He will defend the area around his nest-site against other males but attempts to attract females to the nest to lay their eggs. If he is successful he can then fertilise the brood but must stay and protect them as they develop.

From its simple origins sex has led to an incredible diversity of morphological and behavioural traits. From an individual's point of view it may not always be clear why sex is advantageous, yet sex is now an almost universal phenomenon throughout the plant and animal kingdoms. Environment has had a key role in determining the nature, the direction and the magnitude of differences between the sexes. The structure of a species' habitat affects the size of social groups and influences the mating system. And habitat dictates whether one, both or neither parent must care for the products of their sexual endeavours.

Survival depends on many things: finding food; avoiding predators; and coping with the climate. But sex, and ultimately reproduction, will dictate whether other successes actually pay off. G. E. Hutchinson once described 'the ecological theatre and the evolutionary play'. The theatre dictates the limits and scope of the play, but sex has worked hard on the casting, the wardrobe and the drama which drives the plot.

References and Further Reading

For each chapter texts which include more detailed information and often refer to the examples used in this book are indicated. Where detailed information has been used the source has either been included in the figure and table legends or is indicated in these chapter references.

Chapter 1

Bell, G. (1982). *The Masterpiece of Nature: the evolution and genetics of sexuality*. Croom Helm, London.
Cohen, J. and Massey, B. (1984). *Animal Reproduction: parents making parents*. Studies in Biology Series. Edward Arnold, London.
Daly, M. and Wilson, M. (1978). *Sex, Evolution and Behaviour*. Duxbury Press, N. Scituate, Massachusetts.
Hamilton, W. D. (1980). Sex versus non-sex versus parasite. *Oikos*, **35**, 282–90.
Maynard Smith, J. (1978). *The Evolution of Sex*. Cambridge University Press, Cambridge.
Parker, G. A., Baker, R. R. and Smith, V. G. F. (1972). The origin and evolution of gamete dimorphism and the male-female phenomenon. *Journal of Theoretical Biology*, **36**, 529–53.
Williams, G. C. (1975). *Sex and Evolution*. Princeton University Press, New Jersey.

Chapter 2

Bacci, G. (1965). *Sex Determination*. Pergamon, London.
Bull, J. J. (1983). *The Evolution of Sex Determining Mechanisms*. Benjamin/Cummings, Menlo Park, California.
Charnov, E. L. (1982). *The Theory of Sex Allocation*. Princeton University Press, New Jersey. (Especially Chapter 4.)

Chapter 3

Charnov E. L. (1982). *The Theory of Sex Allocation*. Princeton University Press, New Jersey.

Ghiselin, M. T. (1974). *The Economy of Nature and the Evolution of Sex*. University of California Press, Berkeley.
Piper, J. G., Charlesworth, B. and Charlesworth, D. (1984). A high rate of self fertilisation and increased seed fertility of homostyle primroses. *Nature*, **310**, 50–1.
Tillman, D. L. and Barnes, J. R. (1973). The reproductive biology of the leech *Helobdella stagnalis* (L.) in Utah Lake, Utah. *Freshwater Biology*, **3**, 137–45.

Chapter 4

Fisher, R. A. (1958). *The Genetical Theory of Natural Selection*, 2nd, revised edition. Dover, New York.
Frame, L. H., Malcolm, J. R., Frame, G. W. and van Lawick, H. (1979). Social organisation of African wild dogs (*Lycaon pictus*) on the Serengeti plains, Tanzania 1967–1978. *Zeitschrift für Tierpsychologie*, **50**, 313–26.
Maynard Smith, J. (1978). *The Evolution of Sex*. Cambridge University Press, Cambridge.
Meagher, T. R. (1983). Population biology of *Chamaelirium luteum*, a dioecious lily. II. Mechanisms governing sex ratios. *Evolution*, **35**, 557–67.

Chapter 5

Bateson, P. (Editor) (1983). *Mate Choice*. Cambridge University Press, Cambridge.
Campbell, B. (Editor) (1972). *Sexual Selection and the Descent of Man*. Aldine, Chicago.
Catchpole, C. K. (1980). Sexual selection and the evolution of complex songs among European warblers of the genus *Acrocephalus*. *Behaviour*, **74**, 149–66.
Clutton Brock, T. H., Guinness, F. E. and Albon, S. D. (1982). *Red Deer: behaviour and ecology of two sexes*. Edinburgh University Press, Edinburgh.
Darwin, C. (1871). *The Descent of Man, and Selection in Relation to Sex*. J. Murray, London. Reprinted (1981). Princeton University Press, New Jersey.
Hamilton, W. D. and Zuk, M. (1982). Heritable true fitness and bright birds: a role for parasites? *Science*, **218**, 384–7.
Hatziolos, M. E. and Caldwell, R. L. (1983). Role reversal in courtship in the stomatopod *Pseudosquilla ciliata* (Crustacea). *Animal Behaviour*, **31**, 1077–87.
Krebs, J. R. and Davies, N. B. (1984). *Behavioural Ecology: an evolutionary approach*, 2nd edition. Blackwell Scientific Publications, Oxford.
Queller, D. C. (1983). Sexual selection in an hermaphroditic plant. *Nature*, **305**, 706–7.
Thornhill, R. and Alcock, J. (1983). *The Evolution of Insect Mating Systems*. Harvard University Press, Massachusetts.
Willson, M. F. and Burley, N. (1983). *Mate Choice in Plants*. Princeton University Press, New Jersey.

Chapter 6

American Zoologist (1985). There is a symposium on Paternal Care in volume **25**.

Krebs, J. R. and Davies, N. B. (1984). *Behavioural Ecology: an evolutionary approach*, 2nd edition. Blackwell Scientific Publications, Oxford.

Oring, L. W. (1982). Avian mating systems. *Avian Biology*, **6**, 1–92, ed. by D. Farner, J. King and K. Parkes. Academic Press, London.

Trivers, R. (1985). *Social Evolution*. Benjamin/Cummings, Menlo Park, California.

Index